D1236589

The Ship Asunder

Tom Nancollas

The Ship Asunder

A Maritime History of Britain
in Eleven Vessels

**PARTICULAR
BOOKS**

PARTICULAR BOOKS

UK | USA | Canada | Ireland | Australia
India | New Zealand | South Africa

Particular Books is part of the Penguin Random House
group of companies whose addresses can be found at
global.penguinrandomhouse.com.

First published by Particular Books 2022
001

Set in 10.35/13.8 pt Cardea
Typeset by Jouve (UK), Milton Keynes
based on Text Design by Francisca Monteiro
Printed and bound in Great Britain by Clays Ltd,
Elcograf S.p.A.

The authorized representative in the EEA is
Penguin Random House Ireland, Morrison Chambers,
32 Nassau Street, Dublin D02 YH68

A CIP catalogue record for this book is available
from the British Library

ISBN: 978-0-241-43414-7

www.greenpenguin.co.uk

To Ida

And to the memory of Joseph Alexander Hopkins
1987–2021

'When he first saw a ship rent asunder,
He never beheld one at sea without,
In his mind's eye, at the same instant,
Seeing her skeleton.'

John Ruskin, 1856

Contents

Introduction

Navigating by the spire, I skirt my old school with a rumour in my head: 'From a tradition which the inhabitants have among them . . . a large ship was built near the place where there is now a spring of fine water . . .'[1]

Lying between the Forest of Dean and the River Severn, Lydney can fairly be called an unprepossessing town. I do so affectionately, for I have known it for years. Its buildings straggle and sprawl unpretentiously. It lacks the picturesqueness for which Gloucestershire is better known. Very little you see today predates the nineteenth century, which wracked Lydney with train tracks and industry. It does have a few civic ornaments - Bathurst Park, the Town Hall, the Cross. But it is a place, mostly, where people live, are schooled, work, grow old, pass on. A place that to the outsider would not seem to have much magic. For a while I thought it had no magic either.

Nor did I think it had much to do with the sea. Lydney harbour - for it does possess one - lies at a strange remove from the town proper, as though it were straining for a maritime presence. It is the result of a comparatively recent, nineteenth-century scheme in which the river Lyd was canalised and trammelled down to the Severn. When thriving, it was an artery for the minerals mined from the Dean and sent abroad. I knew it only as a place of truants, mossy quays, stagnant water. But in fact, Lydney has richer, more occluded maritime pedigree than this.

I walk towards the church through a Lydney I have known all my life, a place in which I was raised, confirmed and schooled; a place I later sought to leave and still later shyly revisited. Mostly my recollections of those years are shapeless, but the ensemble of the road, the school and the church vividly uncork all their navigations and humiliations. But I refocus on the spire and come up to the churchyard, looking for the scene of a shipbuilding.

I navigate, too, by Samuel Rudder, aptly named Gloucestershire antiquarian, who walked here three hundred years before me, himself navigating, in turn, by even older voices. In compiling his *History of Gloucestershire*, published in 1779, he heard that 'from a tradition which the inhabitants have among them, the tide [of the Severn] in its usual course formerly came up to a bank of earth called the Turret, just without the churchyard . . .'[2] Now, the churchyard edges an asphalt crossroads, beyond which the land falls gently into an industrial estate where once, according to Rudder's informants, 'a large ship was built . . .' Now, as then, the suggestion seems unreal.

Weekday morning doldrums. Hush hanging over the low warehouses behind steel fences. A mild breeze reaches the crossroads from the estuary, carrying whiffs of turbid water and alluvium, spectral suggestions of the Severn's former presence here. For it was here, somewhere here, that the river once appeared when in spate, and there, over the crossroads and down the shelving road, where that large ship must have been built. It's not an exact piece of triangulation, for the overlain mesh of rails and roads have made the underlying topography quite hard to read. Then I halt upon the threshold of an industrial unit. Within are orderly rows of shipping containers stencilled with the runes of modern freight.

Shortly after this mysterious vessel was built, the Severn heaved its great body further east and left Lydney landlocked. Perhaps this is why, barely a century later, the site of her construction had been lost to memory. To Rudder, the townsfolk of Lydney could relate only imprecise details, but the next century yielded a tantalising clue. During the building of a new railway line to link the town with its now-faraway harbour, the remains of a stone wharf, cannonballs and chain-shot were found somewhere below this concrete hardstanding. Then, before the First World War, certain state papers of the Stuarts, rediscovered by another Gloucestershire antiquary, rekindled the old rumour. And I, casting off for this book, heard the story which made Lydney blaze anew in my eyes.

For here, in the dog-days of the Commonwealth, that

terse, strange period of English history, Lydney and her people built a warship. She was a fifth-rate frigate mounting twenty guns, built to serve in the Commonwealth navy, to patrol, to convoy, to skirmish with Cromwell's enemies. She was a creature of Forest oak and Forest ore, and she projected Lydney into some of the most pivotal naval engagements of the day, extending this parish's horizons beyond its boundaries. She was, in fact, a character who should loom large in Lydney's story. Her name was *Forester*.

This book tells of ships and their shaping of Britain: the lost vessels underfoot, on our seabeds, below our horizons, to which we owe so much. It is a story we have thus far traced back 8,500 years, beginning with these isles' cleavage from the continent. Mysterious vessels duly appeared to make the crossing between the new chalk cliffs. They were delicate things, their yew stitches – no nails, yet – easily sundered in rough weather, leaflike in relation to the vast currents of water they were crossing. And though worked and shaped by bronze tools and the evolving skill of their handlers, there was something definingly organic about these first ships, the sense that, their stitches unlaced, they might return in an instant to their roots.

All this, of course, took place before history began. Only some millennia later, in the first century BC, were the first drops of ink spilled on what Roman scribes called Britannia. Over the next few centuries, foreign hull-forms asserted themselves in these waters, from a very different shipbuilding tradition. Roman warships rode at anchor in the harbours of Britannia which, no longer wholly natural, had been augmented with quaysides and seawalls. The bronze rams on the prows of these vessels were like chisels crafting a new seascape for this annexed province.

Centuries washed by. Roman hulls ceased to work the water, while new, equally aggressive vessels came in arrow-showers from Scandinavia. Viking longships harried the painfully exposed eastern and southern flanks of what had been called Britannia. They were long, serpentine and from an equally foreign shipbuilding tradition to those the Romans displaced. Eventually the harrying ceased with another conqueror, William the Bastard, who crossed the Channel with his men in vessels not terribly dissimilar to those Viking longships – from which they were descended. His flagship was called *Mora*, one of the earliest ships whose name we know.

Trade persisted all the while, carried on in rotund freighters ambling between England, as it was coming to be called, and the Baltic, the Mediterranean. Or skirting England's coasts. Or risking the pirate-infested Channel to bring back Gascon wine, or whatever desirable goods could be attained from the by-now French land mass that was sometimes foe, sometimes friend. Sometimes, with timber fortifications nailed to their frames, they would carry parties of soldiers into the Channel for maritime jousts and battles. Just occasionally these ships would stray further into the Mediterranean and touch the Holy Land. But mostly they pottered in coastal waters, content not to stray too far from the shorelines the first ships knew.

Then, the sixteenth century. The equator, if you like, between the poles of the Conquest and our own time. Near enough on this mid-point, Iberian navigational lore and instruments became available to the English, freeing fleets to travel further abroad. These innovations allowed for transoceanic sailing, land firmly sunk under the horizon, sailors navigating more and more by the stars than by terrestrial contours. Galleons – as we might loosely term these newer vessels – began to supersede their frowzy predecessors. Following centuries of experimentation, they sported three masts instead of one or two. Ships for these voyages had to be lithe and trim, of course, but they also needed roomy holds for New-World trophies, sturdiness for all sorts of seas and strength enough to mount guns. For gunnery was now the order of the day. Gone were the tubby, high-waisted ships of the medieval period, designed for armoured men with swords and longbows; now, these ships' low and lean lines were drawn to the flight of the cannonball.

Russia, Newfoundland, the Far East, the Americas – these new ships expanded the charts of the world, establishing new sites of bounty for European powers. English galleons made fortunes through piracy, poaching upon the transatlantic trades – including slavery – established by Spain and Portugal, fig-leaved into lawfulness by commissions from the Queen. Escalating enmities spawned the Spanish Armada of 1588, a vast fleet of warships driven away by a smaller, leaner English fleet in the Channel. In fast new ships, navigated by French and Iberian teaching, England claimed the seas as its own. Though, to be true, it was as much English weather as English seamanship that seems to have won the day. And, ever since, the English have talked incessantly of both.

Seamanship is synonymous with Elizabethan England, but less so with the following age of Jacobean Britain. This was a strange, uncouth time. Britain, as it became with the accession of James I and VI in 1603, was no longer at war, and naval standards slipped. Without Elizabethan grip, ships were built on the cheap; territorial waters reverted to a state of quasi-medieval lawlessness. Foreign powers warred impudently with one another within sight of the cliffs. No capable navy existed to chase them away. Under the next king, Charles I, ships grew larger and more ornamented. Warships became more spectacular but also more cumbersome, too big to be effective in the bread-and-butter work of convoying merchantmen. Under the Commonwealth, the navy was taken in hand, in a grip almost as tight as that of the Elizabethans. Honed by this zealous dictatorship, the fleet once again fought successful maritime engagements, protected trade and annexed lands.

Now *Forester* was launched, into a story already thousands of years old. In September 1657 she left Lydney for patrols in the English Channel. The next year, she convoyed twenty merchantmen to destinations as diverse as Lisbon and Virginia. The year after, she returned from Copenhagen with her mainmast sprung, was repaired and sent to cruise with Lord Sandwich's fleet in the Mediterranean.

By now, Charles II had been restored to the British

throne. Unlike all his predecessors, the Merry Monarch intimately understood ships and the sea, having spent much of his adolescence fleeing from Parliamentarians in ships of the Royalist navy and indulging, in exile, in the Dutch pastime of yachting. But his relations with the Dutch soon broke down over the question of the trade routes by now firmly established as lines of vast profit between the Old and New Worlds. At the Battle of Lowestoft in June 1665, *Forester* fought in the rear division of the blue squadron, in a sprawling and bloody engagement that revealed just how totally gunnery had come to dominate naval warfare. In lines of battle the Dutch and English warships pounded one another with shot until one side or the other disintegrated.

Ships were lost in mists of shot, but *Forester* seems to have escaped unscathed. That same summer, under Captain Richard Country, she captured a Dutch merchantman off the Norwegian coast. Nations at war, as Britain and the Dutch Republic then were, could lawfully prey on one another's shipping as prizes. *Milkmaid* of Hamburg was bound for Amsterdam from St Kitts, laden with sugar and tobacco. These were the fruits of the plantations, worked by enslaved people, that by this time proliferated across the Caribbean. It is even possible that *Milkmaid* was herself completing the final leg of the 'triangle trade', the last in a series of voyages by which European goods were shipped to Africa, African people were abducted, enslaved and transported to the Americas, and the commodities they were forced to produce were shipped back to Europe.

Under Captain Country, *Forester* took many more prizes in the Baltic and Atlantic: *Fortune* of Hamburg, carrying brandy and town wine from Bordeaux; *St Servane*, carrying beef, hides and herring from Galway to St Malo; *Island of Walcheren*, carrying wine, brandy, tobacco and prunes from Bordeaux; *Mackerel* of Zierikzee, apparently a heavily armed fishing vessel with whom *Forester* exchanged over one hundred shots before she was captured. Then, as now, fishermen are not to be trifled with.

*

I pace out eighty feet on the hardstanding. Pacing halfway back, I sidle five feet sideways to the right, then ten feet to the left. I look up, imagining the height three men would reach if they stood, wobbling, upon my shoulders. Finally, I spread out my arms to make a fathom, and measure two of them. In this way, I sketch *Forester*'s dimensions in a void between the shipping containers, approximately where she was created. Though I gather odd looks from the men in reflective bibs guarding the place, it's a useful exercise in coming to an understanding of her size. And reaching across the void that separates us.

For this is also a search for the lost characters of British history. Ships were protagonists in all the key events of the world until only recently. Think of *Mora*, in which the Conqueror landed at Pevensey. Or the *Matthew*, in which John Cabot made the first English landfall in the Americas. Or *Michael*, the pride of the Scottish navy and the largest ship in the world when she was launched from Newhaven in 1511. Or Francis Drake's *Golden Hind*, the first English vessel to circumnavigate the world. Or *Agamemnon*, built in 1781 at a little Georgian shipyard in Hampshire, the favourite of a certain young Captain Horatio Nelson. These ships were more than mere vehicles. They had individual quirks, soul, discernible personalities. And this is just as true of less famed craft like *Forester*.

To Daniel Furzer, arriving in summer 1656, the Forest of Dean appeared a 'forlorn wilderness'.[3] In some of my less charitable moments I have thought the same thing of Lydney. Yet it was a place, then, where throve the myriad trades required to build a warship; a place capable of building a warship to the tune of £380,000, the modern equivalent of *Forester*'s eventual cost. Furzer was the shipwright who built her. He soon found that Dean timber made for excellent shipbuilding, later lobbying his naval masters to allow him to make masts out of Forest oak, and promoting Forest iron as the best in England. The Forest even yielded a ship's carver, who executed her ornamental work and figurehead; I picture the latter as a verderer of the Dean, replete with horn, staff and green mantle.

We can deduce *Forester*'s appearance from a series of watercolours made by Willem van de Velde the Younger, a Dutch artist who, alongside his father, was commissioned by Charles II to make 'Draughts of Sea Fights . . . for Our particular use'.[4] Van de Velde had begun his career in Amsterdam, depicting *Forester*'s prizes – or mercantile and fishing vessels very like them, anyway – in meticulous shipping scenes. In England, by contrast, he painted warships for the King. As well as the formal paintings of 'Sea Fights', he made many other studies of warships, not of *Forester* but of ships comparable to her.

From these drawings, *Forester* would have had a high stern ornamented with scrolling, gilding and heraldic devices. This higher area of the ship was the quarterdeck, reserved for officers. This was the place from which the ship's direction came and where the whipstaff was located: a lever which turned the rudder. *Forester* was built just a few decades too early to have a ship's wheel, which only appeared in the 1690s, the invention of an unknown genius. The rear or 'mizzen' mast rose through this higher part. Thereafter, the remaining two-thirds of the ship were lower in level, bearing the mainmast in the centre and the foremast towards the bows, which curled together in the curious, ornamental form of a beakhead. Above the waterline, her sides were flanked with nine hatches or ports for the guns. Originally *Forester* carried twenty-two: nine per side, two in the bows and two in the stern. At first, she was worked by a hundred men, but this figure fluctuated depending on how many could be pressed into service or turned over from other ships. Fully laden, she drew twelve feet of water; twelve feet of hull that Van de Velde could not have drawn, since it was hidden below the waterline.

I wish Van de Velde had drawn *Forester*; I wish I did not have to piece her together in my mind's eye from the gleanings in the records. For, at over 300 tonnes, she would have been a powerful wooden presence up close. And somehow, on this concrete hardstanding, she feels maddeningly close, yet just out of grasp.

From the correspondence of Captain Anthony Archer,

her first commander, we gain a glimpse of *Forester*'s quirks. Newly built, Archer reported that she was 'tender-sided'[5] – that is, she made long, slow rolls on the waves and took time to return to the perpendicular. You might picture her making her maiden voyage down the Severn, into the Bristol Channel and round to Plymouth with the long, loping stride of a verderer. Additional ballast seems to have cured her of this roll. Then, in the summer of 1659, Archer reported that her gunports were too low in her sides, so that they were submerged in a gale, and that her stern was too high – though the meaning of this last complaint is less clear.

As well as prize-taking, she passed her life in the grind of convoying. Autumn 1669 found her at Yarmouth, ready to accompany ships to Iceland; the next spring found her in the west dock at Deptford, ready to be graved (her hull cleaned of excrescences). At the Battle of Solebay, in May 1672, she fought in the rear division of the red squadron, again escaping unscathed. Afterwards she was posted to the Mediterranean under Captain Robert Stout where, based in Tangier, she cruised as part of the Mediterranean squadron, convoying merchantmen and attacking Algerine corsairs. In November she sailed for Leghorn (modern Livorno) for two months' provisions. And there she met her end. In the words of Captain Stout's later report to the Admiralty: 'I had not been above a quarter of an hour ashore, till I heard a great report, which shook the house, and immediately one came in and told how the English man-of-war was on fire.'[6] At Livorno, by some unknown accident in her powder magazine, *Forester* was blown asunder.

At the heart of this book is an absence, for ships are definingly perishable things. Sea washes, wears, squashes their hulls. Wind pulls, pushes, prises apart structural members or hull coverings. Salt abrades, corrodes, dissolves until a ship may scarcely be identifiable. Never mind shipwreck or naval engagements. Even in a clock-calm, a ship is a wasting asset.

This is not just a story of ships' lives, but of their afterlives too. And it's hard to overstate how few physical traces there now are of the great ships of the past, how roomy the nation's docks and ports, compared with the maritime ages. These coastal places are like empty sherry casks, drained of a potency yet still fragrant with the scent. Often, only quayside ephemera – a rusting anchor here, a spar or mast there – hint at the rich forms now absent. And if ships could be described as buildings – and indeed they were defined as 'large hollow buildings, made to pass over the sea with sails' in Samuel Johnson's first English dictionary of 1755 – then we have lost whole townships of timber, sailcloth and hemp, great estates of iron.

As it happens, a vessel from the prime of sail does survive. As Nelson's Trafalgar flagship, HMS *Victory* is one of the most famous ships in the world. Launched in 1765, substantially rebuilt in 1801, she illustrates the late pitch of perfection to which sail had been brought by the end of the eighteenth century. Gone is that archaic stepping of deck to the stern – instead, her level decks seem curiously to anticipate those of the later aircraft carriers which can be seen behind her in Portsmouth. Now, she's propped in a dry dock, out of the sea's grasp, fighting not revolutionary France but the more insidious forces of decay.

And by happenstance, also, *Victory* serves as a punctuation mark, for her proposed breakup in the early 1830s – causing outrage, never actioned – coincided with the ascendancy of iron steamships. Hitherto, iron had manifested at sea only in the heavy forms of cannon or anchors; in this context, seaworthy iron ships took a long time to emerge, though they would later make up for lost time with their sheer presence. From the Industrial Revolution came the capacity and expertise that would enable Britain to mass-produce iron on a heroic scale – and what's more, work it deftly enough to shape hulls out of all proportion to anything that had come before. Naval architects, like their terrestrial counterparts, discovered that iron members could be shaped beyond the tree-trunk limits that had previously checked the size of their buildings. Consuming ore instead of oak, these new vessels spared the forests for the first time. When he visited the Forest of Dean in 1802, Nelson ordered a great sowing of oaks to provide for the navy's future. Today, the trees still stand there, unused.

Early on, sail drove some of the smaller iron ships, but only steampower could move them as they grew larger. Fire, for so long nervously proscribed aboard, was now specially accommodated in great boilers and furnaces. First, these engines turned cumbersome paddle-wheels, half-in and half-out of the sea, then wholly submerged propellers that thrummed the ship forward far more effectively. Compared with the thousands of years that had slowly grown the hulls and masts of sailing ships, these new forms appeared seemingly in an instant.

But the greatest change was an unshackling from the wind. Hitherto, seafarers had obsessively courted the breeze. Contrary winds had bottled up fleets in harbours for weeks, while ferocious gales could scatter them in an instant when at sea. Under sail, there was no certainty of passage, no guaranteed course, arrival or departure. Were you to plot sailing voyages on a globe of the world, they would appear as great, looping courses of knots and curves. Steam changed this. Arrivals could be planned to within the day, the hour even; the linear courses of steamships cut like arrows through the looping wakes of their forebears.

Steam advantaged the maintenance of Empire, the flow of people and of international trade, the union of nation states, but it disadvantaged the intricate gaggle of smaller coastal places fringing Britain. For they were not always large enough to accommodate these new ships, which in any case, freed from the vagaries of the wind, could more easily favour specific ports. Gradually, Britain's harbours began to be drained of their contents, just as their quaysides steadily filled with day-trippers seeking atmospheric delights. Amongst we Britons, experience of the working sea, that dirty, malevolent entity, began to wane. By the early twentieth century, it had been repackaged as a commodity to be consumed from harbour towns and transoceanic liners. From the decks of these unequalled steel ships, perhaps comparable only to the greatest, most luxurious hotels on land, the oceans were held at a comfortable remove.

We had transcended wind. We had, it seemed, transcended the oceans themselves; we were shortly to take to the air. Below, the sea which had shaped our history for millennia seemed almost irrelevant, excluded as it was from the ballrooms in the bellies of these vessels. Cruise ships, their successors, maintained this sense of the sea as scenery; container ships, dominating the second half of the twentieth century, brought about a great tidying-up of docksides as cargoes were rationalised and packed into identical oblong containers, marshalled in large new freight ports fenced off from civilians. Slowly, inexorably, our eyes lifted from the horizon.

Today, coastal places are more often drained than brimming. For instance, it's now hard to see how trade with the West Indies made Lyme Regis as cosmopolitan as a capital city. Or from its dilapidated quayside far upriver, how shipwrights at Brockweir in the Wye Valley were once inspired by their Caribbean counterparts to build Bermuda sloops. Or how news – the most precious of all commodities – reached London from the coastal margins, not the other way around. Or how, at the end of the First World War, a great herd of German U-boats surrendered at Harwich, Essex. Like Lydney, so many places once had a *Forester*.

Our relationship with the sea now seems threadbare. Trade is now consolidated – concealed – far offshore in those vast container ships; further inshore, decimated fishing fleets pinprick the North Sea, the Channel, the Atlantic, while dwindled ferry sailings pass in and out of their terminals, maintaining tenuous contact with the islands and the Continent. In many marinas, there are more boats propped out of the water than there are bobbing in it. Shipbuilding has slowed to a trickle and become dramatically skewed: these days, only tiny yachts or nuclear submarines seem to be launched from British slipways.

Although no-one in Britain is very distant from the sea, fewer and fewer people remember how thoroughly its influence

once permeated the country. Who now can recall granite quaysides heaped with fish or hides or spices, forests of masts gently
swaying on harbour swell or clifftop views crowded with ships
of all persuasions? Maritime affairs enriched the country with
striking juxtapositions: oceangoing galleons far upriver, stone
lighthouses far out to sea, precious cargoes in rural backwaters,
broad horizons in narrow valleys.

And yet the sea continues to have a hold over us, a hold of
profound strength despite the fragile or fragmentary remains
of the great ships of the past. In 1856 John Ruskin, mercurial
thinker and conservationist, brooded as intensely on ships and
boats as he had done on anything else: 'But one object there is
still, which I never pass without the renewed wonder of childhood, and that is the bow of a Boat.'[7] He was writing an accom-
paniment to J. M. W. Turner's maritime scenes, *The Harbours
of England*. And he made a striking proposal: 'I should not have
talked of this feeling of mine about a boat, if I had thought it
was mine only; but I believe it to be common to all of us who
are not seamen. With the seaman, wonder changes into fellowship and close affection; but to all landsmen, from youth upwards, the boat remains a piece of enchantment . . .'[8]

Like Ruskin, I'm a building conservationist rather than
a seafarer. And like him, I'm convinced that the sea and its architectures have much to say to us land-dwellers, despite our
frequent ignorance of their ways. I see ships as moving buildings, and this book will explore their fabric, their conservation,
their resurrection. Yet theirs is such a perishable kind of architecture. So little survives of the great ships of the past, because
timber, iron and fibre are inherently vulnerable to the sea and
the salt air. Can we successfully preserve a past which is definingly transient? What may be gained from trying?

If Britain's seafaring history were embodied in a single
ship, she would have a prehistoric prow, be driven by an
ocean liner's propeller, bear a mast plucked from a Victorian
steamship – and all would be outlandishly grafted onto the
hull of a modest fishing vessel. Stone anchors would adorn her

gunwales; painted figureheads would decorate her bows. Archaic props would litter her deck: a medieval ship's trumpet, an Elizabethan captain's chair, a Georgian ship's bell.

Let us call her *Asunder*, a fantastical composite of unlikely parts. Like the maritime history she represents, she would by turns be exotic, improbable, glorious and tragic. And though she may only be a figment of the imagination, the fragments that could build her are real, scattered across the British Isles, some lying quietly in the nation's creeks, others looming prominently over its coasts. In piecing these together, this book will explore what survives of seafaring Britain.

Each of its chapters revolves around a ship-fragment from a specific period of Britain's seafaring history. Each is, symbolically, all that remains of this particular kind of vessel or time. A few of these are in museums, because of their age and rarity, but most are not; I'm most drawn to artefacts which must fend for themselves. And I must acknowledge that nearly all of them are in England – meaning that Scotland occupies less space in this book than its distinctive seafaring heritage deserves. While the locations covered to an extent reflect my own heritage (Cornwall, the Wirral, Gloucestershire and London have all been my hinterlands), the fragments in this book are ultimately chosen through happenstance survival. Through their stories it is possible to see a wider narrative of the ascendancy, decline and fall of seafaring in Britain – as our attitudes to and dependence on the sea begin to change, as the prestige of ships and their coastal berths begins to dim, as the last of the sea drains out of the country, leaving only residue in an emptying vessel.

I come to the end of the track which connects the industrial estate with Lydney Harbour. The Severn at this point is broad enough to be a kind of miniature sea, as good a place as any to launch a whimsical ship. I shield my eyes from the sun and turn downstream, the wind at my back, the broadening, deepening river falling over the horizon. And here I envisage my *Asunder*, riding improbably at anchor, awaiting a favourable breeze. I yearn to see if she swims.

1. A view of an unidentified British fifth-rate frigate, drawn rapidly at sea by Van de Velde circa 1675. Is this how *Forester* looked?

Prow

The Dover Boat

DOVER, SOUTH AND EAST ENGLAND
PREHISTORY, ROMAN AND VIKING/SAXON

> '*Then first did rivers feel upon their backs*
> *Boats of hollowed alder, then the mariner*
> *Grouped and named the stars . . .*'[1]

First through the water, first to touch land, the prow is the foremost part of a ship. And if the nation's ports could be so compared, then Dover might be Britain's prow.

From here, landfall can almost be touched: France, its coastline once the inverse of ours. Eight thousand years ago, an inundation carved this island from that continent, cutting chalk cliffs for the seawalls of both. When new, these cliffs must have resembled gleaming white sashes with dampened hems: finery that was then also fortification, with no breaches through which to pass.

With the passing of the centuries, age has weathered them from new sashes to old curtains, ragged in some places, bunched in others. No longer finery, or indeed fortification. And on the English side, a thin chalk-stream, bubbling up from somewhere far inland, has sliced downwards through the soft chalk-like cheesewire and made an opening.

This - the Dour estuary - is the only gap in the White Cliffs. It is the only landing-place on a twenty-mile stretch of coastline, and it is what drew the earliest seafarers to the future site of Dover. Accordingly, the town has an unmatched maritime pedigree.

I follow the Dour's course in the wake of an ancient prow. Underfoot, a hot ribbon of tarmac; ahead, a prospect of the sea etched with rooflines. I take my first steps through a Dover in suspension: cloudless, windless, unpeopled in the July sunshine. Today, the sky is an undiluted blue that the sea off these shores, in its restlessness, never seems to attain.

Of all a ship's parts, the prow takes the sea's untrammelled force, halving wave after wave to disperse them around the vessel. The town has endured similar pressures. Ever since the Dour carved out a landing-place within convenient reach from France, Dover has been seen as the key to the kingdom, the 'very front door of England'[2] according to the medieval chronicler Matthew Paris. Over two millennia of sea-borne conquest and occupation, fighting and trading, decline and resurgence, the port has parted wave after wave of incomers and dispersed them inland.

It's the hottest day of the year so far, and Dover sizzles. Fortunately, my first port of call here is not the seafront, but the cool shadows of an underpass that leads there. Built as part of a dual carriageway scheme, the walls of this subway celebrate Dover's maritime traditions. Colourful tiles depict the ships it has known since records began. The familiar passenger ferries of today, crisscrossing the Channel to Calais. Early Victorian steamships, strange hybrids equipped with paddle-wheels and chimneys. Sprightly tea clippers. Sturdy, square-rigged warships. An Elizabethan galleon, arched up to the stern like a high-heeled shoe. And then, on one curve of wall, a Viking longboat prowls in the Channel, while a Roman trireme slits the water under the White Cliffs.

But there is an absence on these tiles. For behind this smooth curve of wall, during roadworks here thirty years ago, a remarkable vessel surfaced which rewrote Britain's seafaring history, extending it back by another 1,500 years, beyond the Roman period when written records begin. It lay in an ancient creek with its prow pointing out to sea.

From here, a time-spool back to a drizzly Monday in the early 1990s. They are laying a new road in the south-west periphery, near the seawall: the A20, between Dover and the Channel Tunnel terminal at Folkestone, designed to alleviate the appalling traffic on the existing coastal roads. But the project is nineteen weeks behind schedule and the foreman is under pressure. Today, they are sinking a deep shaft to house a water-pump below the new pedestrian subway and dual carriageway structure. Rain drips onto the archaeologists gathered at the bottom.

This area of Dover is known to have great archaeological potential. It's thought to be where the medieval town wall ran down to the sea, girdling the maritime quarter. The team aren't sure precisely what they'll find, but they have high hopes.

As the digger peels away each layer of earth, the archaeologists inspect the rain-sodden ground and record the finds before they are destroyed by the excavation. Soon they come upon the expected town wall, an impressive masonry structure surviving to a height of nearly four metres underground. Erosive patterning on its south face reveals that this had once been a wall between land and sea; in some areas, the sea had washed it too fiercely, collapsing the masonry and driving shingle into the breaches.

Below the medieval wall they find the massive timbers of a Roman quay, baulks of oak morticed together into square shapes and braced with cross-pieces, a style of construction well known from other Roman harbours (once they had found a winning formula, the Romans were never ones to reinvent the wheel). This quay had been driven into the original bed of the west side of the Dour estuary. So far, the archaeology suggests that this was pristine and undeveloped when the Romans arrived. But this is not to say that the Romans were the first ones here.

Rain patters. With considerable difficulty, they painstakingly record and dismantle the stout Roman quay. Already, the team considers this sequence of finds to be a good haul, with the potential to enrich our understanding of Roman and medieval Dover. But then, from below the Roman levels of the site, emerges a remarkable find that dwarfs the significance of everything else.

They are now perhaps six metres below the modern ground level. In the corner, a dewatering pump works furiously to keep the bottom of the shaft free of the groundwater. Near the pump, an archaeologist sees something. Embedded in the soil is an innocuous-looking side of wood, carved on one face with a semi-circular feature like a section of a doughnut. A strand of what looks like rope trails loosely from it. The archaeologist is intrigued: perhaps a Roman shipwreck? But hasty examination of the surrounding sediments suggests that the timbers lie in undisturbed prehistoric levels. And the way the semi-circular feature has been carved is like no Roman woodwork they have ever seen.

A long halt in the digging; the foreman's heart sinks. He watches the excitement brewing in the bottom of the shaft and feels the delay lengthening. The archaeologists work gently and quickly, peeling back the layers of sediment from the rest of the timbers. What emerges is the midriff of a boat that has been astonishingly well preserved in the moist ground. They find the base and sides of this vessel, hewn with bronze axes from oak, with all the details crisp and comprehensible; they can see ledges for the thwarts, the planks on which the oarsmen sat.

Over the centuries, the vessel had taken the shape of the ancient creek in which it lay, flattened out by the weight of later ages' work above. Freeing the timbers is an arduous task. They are prised from the ground with a variety of improvised tools, from metal excavating implements for the larger pieces, to sharpened lollipop sticks for the last, most delicate parts. And as they are revealed to the air for the first time in 3,500 years, the vessel's timbers begin to glow.

Chalk is fragile when excavated; friable, crushable between thumb and forefinger, yet it endures when compressed underground. The same goes for the timbers of the earliest ships we know.

During the last Ice Age (around 12,000 years ago), most

of what is now Britain,* from the Orkney Isles to London, lay under glaciers. As they retreated, meltwater trickling into the oceans caused a slow rise in sea levels. And with all that glacial weight lifted, Orkney, which was pressed most deeply under the ice, began to rise, while southern England, hardly ice-locked at all, began to sink (the tipping point of this see-saw is apparently somewhere near Hull).

In the Mesolithic period – around 6,500 BC – swathes of this low-lying land were flooded, Britain became an island and what was to become Kent acquired a coastline. Inland, the new island was inhabited by people who hunted, gathered and moved their seasonal settlements on a semi-nomadic basis. It was a time of foraging for roots and berries, and the pursuit of wild beasts such as deer and aurochs, a now-extinct kind of wild cattle. But around 4,000 BC, this restlessness began to fizzle out. People's lives shifted into a pattern that might be recognisable today, defined by agriculture and more perma-nent domesticity. In this new period – the Neolithic – soci-ety started to become more settled and the parameters of life began to expand.

Debate has raged over exactly why and how this change occurred, but the prevailing view is that the new way of life did not originate from within Britain. A recent study of isotopes in Neolithic teeth has shown that many people were born on the Continent. So, this new Stone Age must have arrived by water, but we can only speculate how people crossed the new Channel, for no Neolithic boats have been found in Northern Europe. Yet these vessels must have been skilfully made to carry people and livestock across the future Strait of Dover. One possibility is that they were like coracles, skeletal wooden structures over which waterproofed hides were stretched. None survive.

* For the rest of this chapter I shall use the term 'Britain', though, of course, this name only really begins to apply to these isles from the Union of the Crowns in 1603.

Dated to circa 1,550 BC, the Dover Boat is the earliest known prehistoric vessel in Northern Europe. By this time, thanks to those seminal Neolithic Channel crossings, proto-Kent had been a prospering agricultural area for thousands of years, with widespread evidence for metalworking from the start of the Bronze Age (2,500 BC). Fertile, well-wooded, well-drained and temperate, it is not hard to see why the area was well populated in prehistory, thickly studded with roundhouses and landscapes of sublime monuments.

Until the finding of the Dover Boat, evidence for prehistoric seafaring in Britain was limited to Bronze Age logboats. There are over 150 known examples of these from sites in England and Wales. Shaped into vessels from individual tree-trunks, many of them are skilfully constructed and could have traversed long distances on inland waterways. But logboats aren't seaworthy.

In the early twentieth century, traces of larger, plank-built vessels had been found in Yorkshire, but not enough to reconstruct how they might have looked, sailed or coped with rough seas. The significance of the Dover Boat is that it was found almost intact, except for one end (which lay outside the shaft and was not excavated) and some elements that seemed to have been deliberately removed. As the only substantial prehistoric vessel we have, it serves to embody the astonishingly early origins of maritime Dover - and Britain - after the last Ice Age. These timbers carry with them a sense of seafaring in its first bloom.

I make my way from the underpass, closest to where the boat was found, to the museum where it now resides on permanent display. When they were first unearthed, the ancient timbers glowed golden brown before speedily darkening, the result of a long-postponed courtship between tannins in the wood and ions in the modern air. To avoid any further degradation, the timbers are now kept in a dark gallery, in conditions approximate to the soil.

I want to touch it, but glass intervenes. What immediately strikes me is the vessel's organic character, its soft and frayed edges, the gentle flattening and distressing of the hull giving the whole thing the qualities of a torn leaf. But this is the work of the ground in which it spent millennia, the glacier-like pressing of the strata, the relentless exchange of moisture between the wood and the soil. Enough survives to show that it was once a sturdy and watertight boat, built by people who knew how to contest the waves.

And they were serious about it. To have felled the trees and shaped the timbers alone would have required months of patient, collaborative effort. With remarkable nuance, and much trial and error, the woodwork sections were contrived to fit together without a single mechanical fixing; instead, the timbers were sewn together with fibres of yew – since dubbed 'withies' – and held fast with a series of timber wedges. And, in shaping the lines of the vessel, these boatbuilders saw clearly how to set the grain of the wood against the traits of the water.

What remains of the hull reveals a sophisticated system of interlocking sockets and sections. It's surprisingly large: just over nine metres long and three metres wide in the middle, with room for up to fifteen people. On the bottom, planes of the wood undulate with tool marks, spreading out to the sides where the upwards hull-curves, though slightly crumpled, are clearly recognisable. And at the prow, the hull mysteriously tapers into two pointed sections, making the whole thing strangely reminiscent of a mermaid's purse (a leathery, ten-drilled pouch holding a shark embryo, which beachcombers often find tangled in seaweed).

Originally, the prow seems to have been very like that of a river punt. Experts conjecture that between these pointed sections would have been slotted a third board that rose at a diagonal angle from the base of the hull; as such, the prow would not have curved to a point, but tapered gently to a straight edge. There are strange echoes of this prow-form in the landing craft used during the D-Day invasions of France in 1945, the similarly angled fronts of which fell open to disgorge soldiers upon the Normandy beaches. And perhaps the Dover Boat was a similar kind of people-carrier, albeit without an opening prow.

Over its life it made many voyages, was ground against rocky moorings and hauled effortfully up beaches for safe-keeping. At least, this is the narrative told by the outer faces of the boat, which are much scratched and weathered. But of the nature of its cruises, no trace remains.

Perhaps it coast-hopped, meandering amiably from place to place along the south-east shores of Britain, distributing people, goods or tidings among Bronze Age communities like a floating general store. Perhaps, like a mermaid's purse, its leathery exterior concealed a shark-like purpose, disgorging soldier-sailors to raid rival territories. Or, most tantalisingly of all, perhaps it was the prehistoric equivalent of those P&O ferries that now shuttle between Dover and Calais.

When it had reached the end of its life, the Dover Boat was abandoned in a creek in this estuary. Most of its fittings were stripped back to the hull. And the central section of the prow was prised free and taken elsewhere so that, without it, the vessel could no longer put to sea. It then lay gently decomposing, under the eyes of passers-by, for hundreds of years.

Before Dover was even a glimmer in the eye of a Roman engineer, this place would have been a dramatic portal between the land and the water. Up from where the Dour broadens into the sea, grassy shoulders of chalk climbed steeply into cliffs framing the Strait. Before the A20 was even a shimmer in the mind of a traffic planner, this would have been a resonant junction between the land and the water.

At such places, it is common to find Bronze Age artefacts that have been deliberately broken before disposal. Swords bent in half. Smashed cups. Shattered tools. This is what Dr Francis Pryor, a Bronze Age specialist, discovered in his excavations of sites in the fens of eastern England. The artefacts had been cast into bodies of water from timber causeways that had been built specifically for the purpose. The hypothesis is that these items had been symbolically deactivated before being sent - or returned - to another realm, suggesting that Bronze Age people regarded them as something more than just tools or implements.

Ritually scuttled and left in a creek, the Dover Boat appears to have met a similar end. It must have been a potent object to the crews that had entrusted themselves to the fragile hull and shoved off from land. In the remains of this vessel, perhaps, is carried the embryo of seafaring superstition; in a society which saw the everyday and the spiritual as closely intertwined, the Dover Boat must have carried its passengers far beyond mere points on a map.

And in that confident prow there is a sense of the first launching. A team of sailors and boatbuilders – one and the same, perhaps – dragged this heavy, canoe-like craft down to the water's edge, to the place where the Dour runs out into the Channel. There would have been watchful assessments of the water, since even mildly choppy seas may have been too percussive for the sewn planks. Finally, the boat being pushed into the shallows, the prow broke the waves and nosed towards Normandy, breasting and falling over inshore swells, striking out for the open sea.

Outside, the sun burnishes a pawnbroker's signage. At one stall, under a flutter of gulls, ersatz sausages smoke on a grill; at another, ribbons of technicolour sweets are laid out for inspection; at another, waves of heat-bleached clothes roll across a trestle table. After the subdued conditions of the gallery, the market square overloads my senses. Ahead, along Castle Street, shimmers the green wall of the Eastern Heights, Dover's best vantage point. With some misgivings about its steepness, I begin the sweltering climb. For this is not only a place to feel attuned to the earliest ships, but also to encounter a trace of the first historical fleet that these isles knew.

Another way of describing the Dover Boat might be as an 'aboriginal' vessel from a shipbuilding tradition native to Northern Europe. The prototypical qualities of this boat, together with the home-grown skill that bore it, seem carried in that simple scooplike prow. To take a different example, the serpentine prows of Viking longboats speak clearly of a people who prioritised speed and aggression. When beached, these two vessels would have left tellingly different prow-prints: from the former, a wide, enigmatic mark; from the latter, an indent like an arrowhead.

In 55 BC, a ship appeared off the Dover coast, built along very different lines to the other vessels plying the same waters. She cut through the sea like a blade, long banks of oars on her flanks, each stroke forbiddingly precise. Pulling up and down the coastline, she avoided landfall, keeping out in the Strait as though reluctant to get too close. Vessels of this long, Mediterranean kind had never been seen here before. And after four days of mysteriously examining the coastline, she disappeared.

This, one of the first Roman ships recorded in the Channel, was captained by Volusensus, Julius Caesar's scout. By this time, there had been increasing mercantile contact between Britain and the Roman Republic, which had extended its reach into Gaul (France). But until Volusensus' mission, no Roman had sailed from Gaul for a look at this strange northern island. To Romans the Channel, as part of the Ocean, formed the northernmost limit of the known world; Britain lay beyond, and was viewed with suspicion and trepidation.

Over a thousand years had passed since the Dover Boat was disabled and committed to the creek. Inland, Britain's population had swelled and its settlements were growing. Ramparts were built to fortify natural promontories; swathe by swathe, forests were cleared for field systems. Groups of circular houses had blossomed along their edges, tribes expanded and contracted, and new metals reddened and hissed in their forges. New styles of monument venerated the ancestors, new coinages came into circulation, new styles of pottery proliferated. In this new age, iron supplanted bronze.

Out in the Strait of Dover, crossings had increased in number, but the waters remained relatively roomy and uncongested. Going back and forth were vessels of the same vernacular as the Dover Boat, forging increasingly strong links with traders on the Continent, such that Kent at this time could be described as a 'purely maritime district'.[3] But while the shipbuilding tradition had matured further, maritime Britain was still youthful in character. And, inch-by-inch, the White Cliffs continued to crack and calve into the sea.

Volusensus' foray into the Channel led to two invasions of Britain by Julius Caesar, the Roman general and politician, in 55 and 54 BC. Attempting to land in the Dour estuary, the Romans were repelled and eventually disembarked at a Kentish beach, most probably Walmer. These were abortive affairs, winning Caesar prestige but not much else, best characterised by armour-laden legionaries struggling through the surf. A century later, in AD 43, Britain was again invaded by Roman legions, this time under the Emperor Claudius. Although the location is not certain – Richborough, the Solent and Dover have all been postulated – it was with this seminal landing that Rome finally took possession of Britain.

Roman warships were long, fast and lethal. Depending on their size, they could carry siege engines, artillery pieces and hinged bridges for speedy boarding of other vessels. But their chief feature was their bronze-sheathed prows, for ramming enemy ships under the furious momentum of the oarsmen. And when run in upon the sand to disgorge their soldiers, as they were during the invasions of Britain, they left prow-prints like chisel-marks.

When they had captured an enemy vessel, the Romans had a habit of destroying all but the prow, then using it as a platform from which to deliver victory speeches. 'Rostrum' was their term for the prow, and it was this ancient practice that gave the word its modern definition of a speaker's wooden stage or pulpit. Sometimes these prows were even transported from the faraway frontiers of the Empire to Rome, where they were displayed in the Forum, 'rostrating' the city.

Given that it often best embodies the character of the ship's builders or advertises its purpose, this fetishisation of the prow makes sense. As the point where the curving timbers or sheet iron of the hull come together, it's a natural focal point for artistry, or armament, or both.

*

Under the Romans, Britain's coastlines began to be sculpted. Contrasting with the linear roads and angular towns found elsewhere in the Roman Empire, prehistoric Britain was wonderfully imprecise and natural in appearance, a place of freely composed shapes rather than exactly drawn geometry. For instance, prehistoric round houses were never true circles; they were circular approximations, set out by instinct rather than by measurement. Similarly, very few crisp right angles – in buildings, in walls, in artefacts – have been recovered from prehistoric archaeological sites. And there is little pre-Roman evidence for the sorts of infrastructure – docks, quays, harbour walls – with which the Romans remade the contours of these shores. The harbours were natural, rather than nautical; landfalls were made on beaches, in estuaries.

After the Claudian invasion, Dover – which they named 'Dubris' – was chosen as the base for the Roman naval fleet, and Roman engineers began to develop the Dour estuary. Into the shelving silts of the western riverbank they sunk stout timbers to form a harbour wall, narrowly missing the remains of the Dover Boat. Alongside this, they built a playing-card-shaped fort to the template that recurs throughout the imperial provinces (only this one, uniquely, was a bespoke naval fort rather than a conventional army one). And on the two hills framing the estuary, they constructed a pair of lighthouses.

Standing tall on the Eastern Heights above the town, the surviving lighthouse, or Pharos, couldn't be more geometrical. Its ground plan is that of a square within an octagon. Estimated to have been 24 metres in height when first built, the sides of the lighthouse originally tapered smoothly upwards. With its vanished brother on the Western Heights, the Dover Pharos guided vessels precisely into the new harbour of Dubris. By day, these would have been distinctive landmarks; by night, conspicuous beacons at a time when coasts and countryside were in darkness. They made possible twenty-four-hour Channel crossings, heralding the arrival of a Roman administration unquenched by nightfall.

An exact date of construction for the building has proved

elusive, but it was probably completed in either the late first or early second century AD. By this time, Britain's extant trading networks with the Continent had been streamlined and expanded by its integration into the Empire. To its long-standing exports of tin were added a wide range of other goods; the country was particularly valued for its slaves, dogs and oysters (those from the riverbeds of Richborough, *Rutupiae*, were prized as delicacies in Rome).[4] This traffic was overseen by the *classis Britannica*, the Roman navy, the principal occupation of which seems to have been patrolling and controlling the waters, rather than engaging any rival navies in pitched sea-battles.

Despite the passage of two millennia, the Pharos still stands a remarkable 18 metres high, making it one of the tallest Roman monuments to survive in Britain (although the upper five metres are a later medieval extension). And even after all that time, the Roman ingenuity in masonry is still clearly legible – even courses of flint and rubble bonded with starfish-pink cement, perfectly horizontal tile-courses running in linear red bands around the structure, and regular openings punched in the elevations and crowned with tile voussoirs.

Few lighthouses in the world have witnessed more seafaring activity – every Channel crossing, skirmish or invasion since the Roman Conquest – and it shows: centuries of weathering have left the pebbly stonework looking like batter arrested in the bubbling, particularly where exposed to the prevailing wind. Inside, there is the very British mould-smell of ruins, the cold and vegetal aroma of stonework untouched by sunshine. But light once broke inside here, in the form of glowing embers taken up to fuel the navigational light on the fire-platform above.

That this should be one of only three Roman lighthouses to survive in the world seems to emphasise the depths of Britain's maritime pedigree. And if the Dover Boat revealed the organic virtuosity of the Bronze Age Britons, then the Pharos displays the Roman characteristics of calculated power and physicality. In the absence of any surviving Roman warships or bronze prows, the Pharos serves as a monument to their conquest of the Channel. Sunk permanently into the massing of its stonework are the massed Roman fleets that brought these seaways under Imperial control. It's a powerful expression of a different mindset, one that saw coastlines as there to be worked, the waves of the Strait as there to be ruled, and an island as there to be conquered.

Thistles and other wildflowers gleam on the steep slopes below. Seen from here, on this hot July day, the bay is jewel-blue and feels strangely continental, aptly Latin. Underway to France, two ferries describe lazy arcs in the water. Further off, the continent shimmers in the haze. It's closer to us than you might think; from here, you might imagine one of the first lighthouse keepers, squinting and shielding their eyes with their hands, observing tiny figures across the Channel loading ships, casting off, setting sail.

Hard by the Pharos stands a Saxon church, built nearly a millennium later than the lighthouse, which it once co-opted as a belfry. If the Pharos is distinctively Roman in character, then this much-patched building – its origins uncertain at first glance – evokes the subsequent waves of incomers to Britain, about whom less is known. I duck in there out of the blinding sunshine to be greeted with the unmistakable smell of church: candlemusk, wine dregs and ageing fabric.

In 410 AD, the British Isles ceased to be part of the Roman Empire. By this time, strained to breaking-point by the free flow of barbarian raiders through its provinces, the Western Empire began to sunder. In that same year, for the first time in its history, the Visigoths overwhelmed and sacked Rome itself. Britain was hardly a priority.

The existing population – by then Romano-British after 400 years of Imperial governance – was inundated with migrating peoples from Ireland, France, Germany. Old and new ways of life jostled for prominence. Shedding the buildings and infrastructure of the Romans, these new settlers left different imprints in history and archaeology. For instance, in London, the walled city and quaysides of the Romans fell into disuse, replaced by a 'beach market' – where vessels were simply drawn up on the shore – to the west along the Strand. Most probably the Pharos flickered out.

We know very little about what happened during this time. Some towns were abandoned, while others were redeveloped for mysterious ends. In some places, people appear to have reverted to an Iron Age way of life; in others, the living was newly Germanic. To the west of England and Wales, a form of Roman lifestyle appears to have lingered for longer. In these areas, a number of fifth-and sixth-century stones have inscriptions carved in debased Latin, often alongside inscriptions in the Ogham tongue originating from Ireland. These monuments are the last gasps of a sinking culture.

These 'Dark Ages' (a better term would be 'Early Christian') have been understood as a period of conflict and convulsions as different peoples fought for control of various parts of Britain. An undocumented confusion prevailed until the reign of Alfred the Great (AD 886–899), when the different kingdoms of Britain edged towards unification and the first tentative records of this period, based on imperfect memories, began to be written. At the same time, the Roman infrastructure of harbours and quaysides began to be repaired and expanded; one of Alfred's first moves upon reoccupying the Roman city of London was to re-establish the quays.

But still the seas swirled with turmoil. A new threat arose from the east: Viking raiding-parties, flitting across the North Sea to harry coastal and estuarial settlements.

Around AD 1000, the church of St Mary in Castro was built (or rebuilt – there may have been a forebear) next to the Pharos, during a time when Vikings were exacting tribute from a beleaguered Anglo-Saxon England that relied on paying off the Norsemen rather than defeating them. Sitting in the church, behind the thick flint walls, it's easy enough to imagine the cowering townsfolk counting out the 'danegeld' (as it was known) against reports of Viking longships massing in the Channel. These longships were like showers of arrows into the flanks of a vulnerable country.

Sixty-six years after this church was built, the mercenary chaos in the Channel was brought to a head by the Norman invasion. As with the Romans a millennium previously, a massed fleet appeared off the south coast, this time landing at Pevensey Bay, Sussex. William the Conqueror's fleet consisted of longships which owed their menacing forms to the Norse shipbuilding tradition. Commencing the Norman Conquest and the centuries-long rule of England, it's as if these arrow-prows buried themselves in the sand most deeply of all.

But these soft beaches of the Home Counties, shelving gently into the Channel, offer no evidence today of the landings which so influenced the country's fortunes. There's no sense of a destination hard-won inscribed on the sand, no feeling of a prize taken after an exhausting voyage, no trace of slaughter in the surf. It is only at Dover, where these ribbons of sandy beach meet in the famous chalk seawalls, that these early prow-prints are represented.

At the foot of the Heights, I grip a cold pint in a pub where the internal walls are scribbled in marker pen with the names, dates and times of those who have swum the Channel. Just imbibing the town's deep history is similarly gruelling, I think, before spotting an inscription from three elderly churchwardens, who had swum there and back in a mere twenty-one hours. Chastened, I take deep draughts, trying to imagine such a crossing unshielded by any hull.

Wood is tactile and warm to the touch, whether at the prow of a dinghy or a longship. I had been curious to see whether the Dover Boat still held any warmth in its ancient timbers, but there was no way of running my fingers along their grain: the climate preserving it is too delicately balanced for prying hands. Only specialists will ever now have the privilege of touching the boat, but perhaps it is right – in the interests of both mystique and conservation – that some things are withheld from the public grasp.

Contrastingly, in the nave of the church, I sat on a wooden pew as cool and smooth underhand (I imagine) as a longship's thwart. After 2,000 years of weathering, the surfaces of the Pharos were wonderfully textural: hot flint-shards, abrasive mortar, nubbly stonework. The town is full of textures and surfaces like these, deriving, in part, from the shelling it endured during the Second World War. While many fine historic frontages survive, they are interspersed with shoddier modern buildings on plots destroyed by cross-Channel gunnery. Cheap new elevations are juxtaposed with venerable masonry, slapdash repair and neglect. Consequently, Dover feels like one of the most tactile places I know.

And aptly so, for people have been reaching out to it for thousands of years. France's closeness is felt in many ways here, not just in the havoc wrought during the World Wars, or the scribbled triumphs of cross-Channel swimmers. Smooth and faintly distorted, the town's surfaces – walls, pavements, carriageways – collectively have the feel of a stone threshold step trodden to a scoop. For thousands of years, Dover has been one of the main entry-points to Britain. Yet, traipsing around the town today, there is the distinct sense that it's now a port with waning importance, that the waves it greets are subsiding. Although much still comes through Dover, much else that it would once have processed now passes through the air, to other ports, or is freighted under the Channel.

But in the calm of the harbour, sheltered by massive concrete breakwaters, there is evidence of a booming and more desperate kind of traffic. Here drift abandoned migrant vessels: pinpricks of colour that bob in the sea lanes, weaving precariously between colossal tanker hulls and laden almost to sinking with life-jacketed passengers. They shove off from the French coastline in delicate dinghies, brittle catamarans or weary trawlers. Unlike the painstakingly shaped vessels of early history, these small boats are mass-produced to global rather than local templates. They are cheap, anonymous and could have come from anywhere.

This is seafaring stripped back to its basics, by people who don't want to be seafarers at all. In 2020, 5,000 people attempted to reach Britain from France by sea[5] – a dramatic upsurge in this type of crossing. Their vessels are caught mid-Channel by the authorities, or sighted inshore, in difficulty, or found crumpled against the breakwaters.

Before I get the train to London, I see a grey dinghy abandoned on a slipway. I have no idea whom it carried, or under what circumstances. But there is something human-like in its squashy hull, firm when launched, pliable when deflated. It makes for a heady continuity with the prows this place has known over the past 4000 years, all of which were carefully shaped to carry, to indent, to attack. But it's enough for this lone prow just to touch the land.

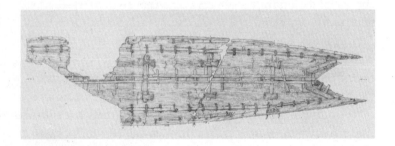

2. The Dover Boat as found: archaeologist's drawing of the interior of the vessel and the 'swallowtail' prow.

Trumpet

The Billingsgate Trumpet

CITY OF LONDON; WINCHELSEA, EAST SUSSEX
MEDIEVAL

Winter solstice, far removed from the slipway. In Dover, I had walked in the summer equivalent, a day of infinite width; today, we walk the north edge of the Thames in a brief, deep crack of sunlight. Ringing in my ears are the words to a carol: 'I saw three ships come sailing in / on Christmas day, on Christmas day'. I hum it brokenly as we walk from Blackfriars to Billingsgate.

At Christmas, as if through a pinhole in a dark screen, there comes a faint projection of medieval England. To offset the long nights we don, like a heavy old robe, the rich colours and superstitions of those post-Conquest centuries; for a few days, we gorge like those kings (wearing crumpled imitations of their crowns); for a few days, we recall the saints firmly embedded in the medieval calendar. At all other times of the year, this period in English history can feel as remote as the Holy Land.

Spanning the period from the Norman Conquest of 1066 until the Tudor ascendancy of 1485,* the Middle Ages are today commemorated in Britain - and, perhaps, caricatured - by a plethora of surviving castles, timber-framed buildings, town defences, wayside crosses, churches and other features of the townscape and landscape. Fortunately, these monuments are like anchorages, helping us to make sense of the period. In the popular imagination, they evoke such feudal characters as knights, maidens, peasants and clerics - but rarely sailors.

Of the medieval seascape there are precious few traces.

---

* There is no general agreement on the extent of the 'Middle Ages' or medieval era. I cleave to a rather old-fashioned definition here: the periods formerly known as 'Dark Ages' and 'Saxon' are now often considered 'early medieval'.

At Smallhythe, Kent, an oblong pond on a grassy estate may once have been a medieval dry dock in which ships were repaired and constructed; at Newlyn, Cornwall, and Lyme Regis, Dorset, small harbour arms extend tentatively into the sea, medieval in outline if not wholly in stonework (having been much rebuilt); single-masted silhouettes flutter on the flags of Scarborough, Hastings and certain other ancient sea-towns.

Yet this was a time when seagoing trade was expanding and consolidating; when fish was central to the diet of an expanding population; when ships were crucial vessels for diplomacy, royal marriages and crusaders; when England itself was in maturation, coalescing into the kingdom we recognise today. Maritime affairs were assuming a background familiarity in many lives; England was conversant with the sea but not yet ready for the ocean.

It might be described as the colourful adolescence before a coming of age. And like most adolescences, this was a time of wobbly navigation, occasional bravado, false starts and thrilling advances. The first notable naval battle of the period was between the English state and a pirate; the naval battles themselves took the form of terrestrial-style swordplay between armoured men surreally transplanted to swaying decks. In twelfth-century London, foreign crews would sing a Greek hymn as they sailed up the Thames. And, most eye-catchingly, curious figures strode the decks of ships, wielding gleaming, long-stemmed trumpets.

Now, the city's docksides are sunk beneath centuries of later development, its harbours silted up and long abandoned, its ships entombed in beaches or riverbeds. But the medieval shipscape is not altogether irrecoverable, if you know where to look.

Of brass, once gleaming but now dulled to a warm gold shade, it runs for over two metres, slender-stemmed then gradually flaring, from the mouthpiece to the sounding end. Ornate knobs and ribbed mouldings glimmer along its length. Eight centuries of existence have lent the brasswork a nicked, malleable quality, as though it could be warped by over-zealous fingers.

Most objects merely come from a time, while others seem to enshrine the time and the place from which they originate. Found in 1984 during excavations of the waterfront neighbour-ing Billingsgate Market in the City of London, the Billingsgate Trumpet, as it is now known, is the only complete medieval straight trumpet (or 'buisine') found in Northern Europe. And as the only known example of a ship's trumpet to survive, it makes real an element of the medieval ship that had hitherto only been a matter for speculation.

Yet in the Museum of London, where it now lives, this status seems curiously downplayed. It glints in a glass mon-olith, surrounded by debris scooped from the riverbed: bent coins, a broken anchor, a splintered oar. On the first of my many trips there (I work in an office around the corner, slip-ping out at lunchtimes to see it), I studied the relic in a drift of visitors who seemed oblivious to its significance.

Later I found that the language of trumpetry, with its pungent words and obscure meanings, equals that of seafar-ing for its power to evoke. The trumpet's sections are called 'yards' – also the name, on a ship, for the cross-pieces on the mast from which sails are hung. The sound-emitting end is called the bell, shaped upon an anvil-like tool known as a man-drel by the instrument-maker hammering a brass sheet into the desired conical form. The bell's edges are delicately fused into a meander seam. And, just as good, the study of musical instruments is known as organology.

Until it surfaced, the ship's trumpet was an elusive pres-ence in the historical record. Here and there, it appears in the margins of illuminated manuscripts, or on the wax seals of medieval sea-towns, or passingly observed by medieval chron-iclers. Equally elusive are its origins: scholars are unsure just how such an exotic instrument came to be wielded upon the decks of English ships. And it was elusive, too, in its disappear-ance: at some unknown point in the transition from seafaring adolescence to maturity, the trumpet seems to have been ousted by the smaller, more portable boatswain's pipe.

At first glance, it does seem an unlikely ship's tool. Many depictions of it are religious or chivalric in character: banner-hung instruments blown by heralds at jousts, or by angels in the margins of heaven. Unlike the prow, its form and features are not obviously shaped by the sea. Indeed, on a practical level it seems not to belong there at all, looking too unwieldly in length for the cramped confines of a ship. And its extravagantly ceremonial character feels at odds with the simple practicality of other nautical paraphernalia.

Yet it had many uses onboard. To Peter Marsden, a former archaeologist of the City of London and maritime history expert, the instrument would have been blown by a medieval skipper standing at the stern of the ship as a means of giving orders to the crew as they went aloft to set sail.[1] Designed to project a small range of sounds over great distances (rather than for playing intricate tunes), it would have carried signals over the roar of the surf to other vessels and even, when in range, to stations on the coasts. It would have transmitted orders to the crew, cutting decisively through the blizzard of small noises – shouts, squeaks, rustles, creaks – characteristic of a sailing ship underway. And it would have blown a path for its vessel through congested estuaries and bustling ports.

If, with its seashell-like curves, the prow is the closest to nature a man-made thing may ever get, then this superbly worked trumpet is its opposite. And if the prow captures the moment when the simple problem of seaworthiness was solved, then this trumpet embodies a subsequent phase, rich yet ambivalent, when England was working out how best to sail.

Consider the Shipman, tanned, plain-gowned and circumspect. A dagger dangles on a lanyard from his neck, freely exposed to view; while his mercantile passengers sleep, he draws off cups of their wine for himself. He knows all about tides, currents, moons, harbours and the wiles of local pilots, more so, it is said, than any other skipper on the Atlantic coasts. And he knows all the havens, from Gottland to Finisterre; every creek from Brittany to Spain.

This is Geoffrey Chaucer's Shipman, one of the fictional pilgrims in the Canterbury Tales. Chaucer knew what he was talking about: as a frequent traveller to the Continent on royal and diplomatic business, and later a Controller of Customs of the Port of London, he mixed regularly with skippers of the Shipman's type. A portrait penned probably in the 1360s, this is perhaps our most vivid depiction of a medieval skipper. And he embodies the limitations of the ships of this period.

To begin with, most medieval vessels crept tentatively from creek to creek, from haven to haven. Navigational technique went no further than first-hand experience and received tradition. Ships, however sturdily built, were single-masted and awkward to handle. Most could not cope with the deeper ocean. Perhaps more tellingly, it was not a priority to voyage in that direction. As the extent of the Shipman's knowledge implies, ships rarely strayed beyond a network of sea routes between the Baltic ports, the Channel shores and the Atlantic seaboard. When they did, the priority was the Holy Land.

Here and there, where modern redevelopments block the frontage, we are forced up into a tangle of older, neglected streets which were once arteries between the markets of the City and the Thames. Although I've worked in the City for a while now, I somehow lead us astray and find our way blocked by the bleak canyon of Lower Thames Street, which was widened in the 1960s and severed these medieval arteries in two.

So, we stray from the river for a while. On Josa's (my wife's) suggestion, we climb up into the City to find a better approach back down towards the Thames. Here we stumble across St Nicholas Cole Abbey, a deconsecrated church with a bulgy spire like the end of a bugle. Happily, it's dedicated to the patron saint of seafarers, although the relevance to them of this fourth-century Bishop of Myra seems tangential (apparently, he saved some sailors from drowning, but was no mariner himself).

In London, by the end of the twelfth century, there stood over 120 churches dedicated to a vast array of saints, from obscure Turkish bishops to martyred Saxon converts. So great a number reflects both the overwhelming piety of the medieval city and, brimming within the old Roman walls, its sheer density of occupation. In addition to the churches, the city was further enriched by the rambling complexes of monasteries, nunneries and friaries, each with their own distinguishing liveries and liturgies. And above them all towered the spire of old St Paul's, then the largest cathedral in England.

Unless ground conditions or neighbouring buildings forbade it, these churches would always be aligned with their altars and chancels facing east. Stained glass depicted a wide range of biblical scenes and usually some representation, however stylised, of Bethlehem, Nazareth, Jerusalem and other sacred destinations. Near the riverfront to the west of the city, the church of the Knights Templar was a literal representation of the holiest church in Christendom. With its unusual circular nave modelled on that of the Church of the Holy Sepulchre, it was consecrated in 1185 by Patriarch Heraclius of Jerusalem and still nestles exotically in the capital.

Saturated in Catholic teachings and rituals, in almost constant psychological alignment with the Holy Land, it's unsurprising that many felt the pull of pilgrimage. Their destinations were various, chosen according to time, wealth, fitness, sanity or allegiance towards a particular saint. Some felt British shrines like Walsingham, Norfolk, or Canterbury satiated the pull. Others went further afield, to continental shrines such as those at Rome or Santiago de Compostela. Waterfront excavations in the City have revealed a vast quantity of pilgrim souvenirs from 39 British and 109 European shrines, ranging in geographical spread from Sweden to Bari, above the heel of Italy. But the ultimate journey was to Jerusalem itself.

To go there was to acquire impressions of the Holy Land more vivid than the brightest stained glass. An early twelfth-century account left by a Saewulf, a British pilgrim, vividly sketches the surprising number of devotees then streaming into Jerusalem following its recapture in 1099 by European princes during the First Crusade. It was a nightmarish journey, of the sort that could only be driven by intense piety, alternating between arduous, thief-infested overland routes and perilous, storm-battered passages by sea. Such were the dangers that knightly orders, the most famous being the Templars, emerged to (lucratively) shepherd these helpless pilgrims.

By far the most prestigious form of pilgrimage was to go on crusade to defend or extend the Christian presence in the Holy Land. And it was for crusade that in April 1190, amidst the shriek of instruments and riot of coloured banners, a fleet set sail from Dartmouth. Its voyage south past the French and Portuguese coasts, through the Pillars of Hercules and into the waters of the Mediterranean illustrates, on a grand scale, the tentative, coast-hugging character of medieval seafaring.

Mustered by King Richard I to carry troops and provisions for the Third Crusade, this fleet was a sizeable one. In contrast to the cash-strapped efforts of previous monarchs, Richard had amassed a huge war chest to fund this campaign, reputedly (and almost certainly apocryphally) claiming that even London would have been up for sale had there been a buyer. And not only was this crusade well resourced, but it was well commanded. Earning the epithet 'Lionheart' during his lifetime, Richard was a shrewd and battle-hardened king. Although the crusade would fail in its main objective, the recapture of Jerusalem from the Sultan Saladin, other successes lay ahead: the capture of the strategically vital island of Cyprus, the recapture of the city of Acre and the conquest of coastal territory from Acre to Ascalon.

Jostling in the mouth of the Dart, their limp sails suddenly wind-stretched, the ships themselves were still of the Scandinavian tradition. Formed of thick, overlapping planks (or 'clinker-built', in shipbuilding parlance), the long hulls of these vessels had, in design, barely moved on from the Norman longships of 1066 or even the Viking longboats that had preceded them. Above the waterline, however, there had been some advancements to the layout: the mast had been set amidships to better catch the wind, and small timber castles –variously used as shelter, watchtowers and fighting platforms – had been added at each end.

Laden down with weapons, soldiers and horses, it was to be a long, halting cruise. By this time, on account of Henry II's (Richard's predecessor) marriage to Eleanor of Aquitaine, the English kingdom stretched across the western half of France, down to the Pyrenees; accordingly, Richard's fleet hopped along the Atlantic seaboard, swelling in number as it revictualled at various ports along the way. But the King himself was not among them. In December 1189, Richard had crossed from Dover to Calais and made his way overland to Vézelay, a prestigious monastic complex on the border between the realms, where he met the French King Philip, the crusade's co-commander, to agree a joint strategy and the shares of the spoil. Along the way, he forged alliances, obtained more ships for the fleet and arranged the government in his absence.

In high summer he arrived at Marseilles to rendezvous with the fleet after it had passed under Spain and into the Mediterranean. But, unbeknownst to him, the fleet was running late: it had invaded Lisbon and spent days rioting and plundering the town. After waiting fruitlessly for a week, Richard set sail for Sicily, the next rendezvous point, and coast-hopped at leisure between Italian ports.

Eventually the parts of Richard's fleet came together at Messina. On 10 April 1191, almost exactly a year after leaving Dartmouth, they finally set sail for Outremer, as the crusader settlements in Palestine were then known. According to contemporary estimates (almost certainly exaggerated), there were about 180 ships of the Scandinavian type and 39 galleys hired in Italy - fast, oar-pulled vessels descended from the Mediterranean shipbuilding tradition - organised in eight divisions. Richard himself commanded one of the galleys. And as they made their way over the Mediterranean, the ships huddled close, ordered to not stray further from one another than the range of a trumpet blast.

Crusade and pilgrimage strengthened linkages between northern Europe and the eastern Mediterranean. And around the time that the crusades began, trumpets resembling the one found at Billingsgate began to appear in European art. Arabic influence is shown in the decorative knobs along its length, grafted onto a straight-stemmed form of Byzantine origin. Although we cannot be certain, it seems highly probable that returning crusader fleets carried the archetype into Europe, whence it was honed and replicated by the brassworkers of Nuremberg and Paris.*

No home-grown instrument, then, the ship's trumpet, but one that originated in the Holy Land. It embodies a peculiar crossover between the prosaic business of ship-signalling and the potent symbolism of the crusade. And as the only surviving example of its kind, the Billingsgate Trumpet powerfully commemorates the furthest from England a medieval ship would go, limited by seaworthiness, circumscribed by piety.

There is a kind of symbiosis, I think, between the trumpet and the Thames. Sometime around 1300, it was dropped overboard from a ship docked at Billingsgate. For eight centuries, as continual rebuilding and expansion gradually erased the medieval waterfront, the river mud kept this relic safe, revealing it again only when a metal detector swept a patch of newly cleared earth. Now, the trumpet hints at the river's early richness in traffic and trade.

From an alley mouth we look out over the Thames, enjoying the sharp contrast between its breadth and the narrowness of our vantage. The water ripples inanely, like a screensaver. Few vessels pass by, and not just because it's nearly Christmas;

* An English trumpet-maker working in Paris, Rog. L'Englois, has been tantalisingly cited as a possible creator of the Billingsgate Trumpet by the musicologist Sabine Klaus. See John Schofield et al., *London's Waterfront 100 to 1666* (2018) for the essential account of the excavations and the interpretation of the finds.

there are fewer ships on the river than there once were. Under a layer of waterproofs, our newborn daughter Ida sleeps soundly on my chest; I wonder idly what she'll make of the Thames. Somewhere a church bell tolls twelve times, propelling our footsteps onwards.

Rivers were once hugely important. On a well-known, early thirteenth-century map of England, Wales and Scotland drawn by Matthew Paris, they are by far the most prominent topographical features. At a time when the road network was primitive and at times impassable, rivers provided fluid transportation for medieval people and their cargoes. In the fourteenth century, lugging goods overland was, on average, twice as expensive as gliding them downriver.[2]

Settlements thrived along riverbanks and bends; something like three-quarters of all medieval fairs (not just social events, as they are today, but highly important commemorative, commercial occasions) were held near a navigable river. And their deep estuaries provided sheltered anchorages for England's merchant and naval fleets and the shipyards that built and maintained them. With their narrow beginnings and splayed ends, you might think of England's rivers as trumpet-like, blowing a potency into the world. And the Thames is a great, crumpled exemplar.

Founded by the Romans in the first century AD, London's position at a conveniently fordable part of the Thames has never fallen out of favour, apart from the Saxon initiative to hold, for a few centuries, a beach market further west along the riverbank. From the ninth century, after the walled city of the Romans had been reoccupied, the quays bustled with traffic despite frequent Viking raids. An ordinance of the Saxon King Ethelred (c.990s) reveals that, even at this early date, continental merchants were landing various cargoes at Billingsgate.

Accordingly, when the Norman fleets sailed up the Thames shortly after the Conquest of 1066, they docked at a port of London that was already prospering. Through international trade, Londoners had grown wealthy and influential, their city the most cosmopolitan in the nation, not to be steamrollered by Norman administration like other towns. William merely raised fortifications to the west and east of the City as a reminder of his power, having issued a charter to the Londoners confirming their 'ancient rights and privileges'. The city continued to thrive.

By the 1170s, when London had become de facto capital of an empire that stretched all the way to the Pyrenees, the city could be described as 'one whose renown is more widespread, whose money and merchandize go further afield, and which stands head and shoulders above the others'.[3] In this description of London that prefaced his biography of Thomas Becket, the cleric William Fitzstephen described how 'merchants from every nation under heaven are pleased to bring to the city ships full of merchandize'.

Barely contained by its walls, the medieval City was a dense thicket of timber gables, stone towers and iron finials. Guilds of all trades held sway, from Mercers to Merchant Taylors, from Poulterers to Pepperers, their freemen sauntering with prestige about the place. Between the marketplaces clopped hooves, trundled wheels, squelched feet, while Latin, Saxon and Germanic tongues jostled in the meetingplaces. Gossip, fire and occasional plague leapt from house to house, street to street while, between the wharves and the countinghouses, profits were skimmed from boatloads of wool, furs, wine and spices. Ordinances from the City fathers, sternly issued and reissued, only just held this chaos in check.

Finished circa 1209, old London Bridge marched spectacularly over the river on nineteen stone footings, a line of shops and houses tottering above, barring all but the smallest rivercraft upstream and creating a focus of shipping at downstream Billingsgate. As if in recognition of its mercantile importance, the riverfront was constantly being smoothed and modified throughout the Middle Ages. Carrying on work begun by the Romans, the wavy shoreline was straightened first with timber revetments, then with stone river walls, enlarging into the river both the footprint of the city and its capacity for goods.

Today, there is only one point along the riverfront where a

sense of these wharves and landing-places might still be gained. Queenhythe, like Billingsgate, was founded by the Saxons in the ninth century and remained in active use until the twentieth. Positioned just upstream from Southwark Bridge, it maintains the medieval form of a square inlet, breaking the modern linearity of the embankment. Over the many centuries of its operation, everything from fish to flour, tin to trinkets, passed through this small frame of water. Still tidal at this point, the river fills and withdraws from it twice a day, unveiling a foreshore from which, I suspect, handfuls of potsherds, glass spicules and other granular relics might be prised.

Such wharves were the destination for wares from Gascony, Flanders, the Rhineland, Italy, Spain and Mediterranean ports, through which were transhipped goods from even further afield. Foreign merchants had their own enclaves in the City: those from the Baltic were especially influential and operated from their own fortified complex near Dowgate. Near the Tower, by the end of the fourteenth century, Genoese galleymen operated from their own quay.

Ancient custom governed the import, landing and distribution of produce. A law of c.1130 governing the trade of the merchants of Lotharingia (now the Low Countries) is a fascinating illustration of a lost mindset. Once a year, a wine-fleet set forth from Lotharingia for London. Upon arriving in the Thames Estuary, the law required their crews to arrange themselves in formation on deck, raise their ensign and (if they wished) sing the 'Kyrie elesion', the Greek prayer of praise and thanksgiving, until they reached London Bridge. This peculiar stipulation seems to have originated in an earlier Saxon law to weed out pagan crews; by this time, it seems to have become something of a tradition. When the boats were finally docked and unloaded, the King had first refusal on their cargo, for a period of two ebb tides and a flood, followed by various nobles in descending order of seniority.

Wine was the chief import, and wool the main export; both were carried back and forth over the Channel in big-bellied, high-waisted ships known as cogs. Developed by Frisian shipwrights, they were the commonplace freighters of the age and by the thirteenth century had become ubiquitous on the trade routes of the Channel and the Baltic.

Such was the demand for English wool that portions of the fortunes it created could be donated to public-spirited building projects. A number of England's medieval bridges were said to be built upon sacks of wool, alluding to the method of their financing rather than their actual foundations. In places such as Norfolk, a county which grew immensely rich on the wool trade, there still stand the churches which merchants built or enlarged to dizzyingly elaborate effect, such was their desire for absolution. And the profits from wool were so extensive that, for a time, warmongering monarchs appropriated its revenues to wage continental campaigns.

In turn, the profits from wool and its subsidiary trades in cloths, dyes and other accoutrements wove themselves into the tapestry of medieval life. At Billingsgate, the excavations that produced the trumpet supplied a vast stream of pottery, buckles, cutlery, glassware and other accessories, revealing that London's consumer culture, presumed to have blossomed in early modern times, had much deeper roots indeed.

Climbing the stairs from the riverfront up to London Bridge, we stand in the middle and look both ways; downstream, through the picture-frame form of Tower Bridge, until the water is lost to sight as it bends around towards Millwall; upstream, back at the route we have taken along the Thames path. These days, the riverfront runs smoothly through the city, shaved of its rabble of jetties and wharves. And from this midriver perspective, the city glitters like a dashboard.

Although London is still tactile in places – the pale stone spires, the brick back walls, the quaint terracotta frontages – much was lost to German bombs and much else has since been sacrificed for the city to stay globally pre-eminent. In some of its old wards, smooth planes of glass outrun the weathered lengths of older walling. Much of the street pattern remains medieval in origin, but these narrow, slotlike streets have become ancient frames for modern views. New towers rise far above the antique parapets and cornices of the historic city, their irregular shapes glinting like a glassblower's discards in the right kind of sunlight.

Of course, a historic, human scale of building still lingers in places. The city retains shards of traditional townscape as tactile as anything in Dover. Yet the fine frontages of tooled stone and sculpted brick, the faces to the world of merchants' halls and houses, churches, corporate headquarters and civic buildings increasingly compete with the gloss of new glass, lifeless to the touch. The material seems to emphasise the City's new abstraction, especially on the grand scale required for global offices.

And for the clippers shooting commuters between Battersea and Canary Wharf the river now runs emptily. Commerce still flows in the city's thoroughfares, but the commodities now traded are of the intangible kind, heaped not upon wharves but in server banks. To the roll-call of ancient guilds that begins with the Mercers, Grocers and Drapers have been appended new guilds of Information Technologists, Management Consultants, International Bankers. Aside from people, the only commodity now to leave the City by river is its rubbish, loaded onto barges near Cannon Street Station.

Imagining this view bridgeless and choked with Hanseatic cogs, Genoese galleys, Thames shouts* and everything in between, it's easy to see how useful the trumpet must have been in clearing a path through the traffic. Its brash note would have been an unmistakable command to get out of the way. And unlike the ships' horns of today, mechanized and buried somewhere in the fabric of the vessel, the ship's trumpeter, wielding his glittering instrument, would have been a commanding spectacle.

Literally so, for the trumpet was used to command. Confirmation of this appears on the insignia of medieval common seals – for sealing documents with wax to secure them – of medieval ports from Great Yarmouth to Dover. Unlikely as it may

* A flat-bottomed medieval river boat.

seem, these administrative paraphernalia are some of the most evocative images of medieval seafaring to have survived. They have a simple, dream-like quality. Under stars and moons, aboard ships exaggeratedly curved and proportioned, the crew clamber in the rigging or haul ropes upon the deck while the trumpeters wield their long instruments at the stern.

3. The evocative medieval seal of Winchelsea showing a medieval ship with trumpeters at the stern

Until the sixteenth century, Billingsgate was a marketplace for various merchandise – corn, coal, iron, wine, salt, pottery among them – but thereafter was exclusively for the sale of fish. Until the mid-nineteenth century, like Queenhythe, it took the form of a little indent in the riverfront; this was infilled and in 1876 the City opened a fine market building in the

French Renaissance manner. It lasted barely a century before the market relocated to the Docklands. Now, at Billingsgate, the lifeless market building sadly regards the Thames, set back from the river by a modern plaza, spotless and regrettably unfishy.

It was here that the trumpet was dropped into the river through the gap between a tilting deck and the level dockside, a loss perhaps noticed straight away by its hapless owner, or an absence only realised, to curses, much later on. Josa imagines the sinking yards of the instrument glinting one last time before disappearing for good; she pictures them glinting again, eight centuries later, eager hands having scraped away the river earth, in the daylight of a very different age.

Metallurgic analysis of the trumpet's four sections has shown that they come from different instruments and vary in quality. The bell and third yards originate together, while the mouth yard and the second yard are later substitutions. For workmanship, the finest is undoubtedly the bell yard, while the second yard is the most inferior, being little more than a poorly wrought length of pipe. Throughout the whole instrument there are repairs of varying care, some finely done, others hastily applied.

Various conclusions might be drawn from this. One is that, by the time of its loss, the trumpet's owner had suffered a few reversals of fortune. Or maybe it's the case that, once they had lost their initial sheen, and been repaired once or twice, these sorts of trumpets were demoted from prestigious uses on land and turned over to a more workaday life at sea. Or perhaps ships' trumpeters are so elusive in the historical record because they were itinerant, never fully part of a crew, making their way from ship to ship and not earning quite enough for the upkeep of their fine instruments. Quite typically for the study of the Middle Ages, we may never know the full story. The projection into our time of this strange period, of its exotic, tarot-like characters, can be faint and flickering.

Ida twitches and wakes. For an instant, her uncomprehending eyes flit and slide across the whole view, then light upon the river's brown breadth.

One of the fascinations of the ship's trumpet is that it is not a definingly nautical object. On land, trumpets of this kind had heraldic significance and courtly functions. They initiated jousts, announced dignitaries, propelled knights into battle. Although it formed a crucial part of medieval ships' equipment, the instrument had roles in both realms. In this, it reflects the embryonic quality of the medieval navy, in which the tactics of the land were awkwardly transposed onto the conditions of the sea.

From the longship-like Scandinavian hulls to the high-waisted, cargo-friendly cogs, a characteristic of medieval ships was their interchangeability between times of war and peace. No such thing as a state navy yet existed in England. A peculiarity for an island nation, perhaps, but in the centuries following the Norman Conquest there had not been much demand for battleships.

England continued to control the coasts of the Channel and the Atlantic seaboard down to the Pyrenees until the reign of King John (1199–1216). The prospect of naval warfare in these seas was remote. But John lost much continental territory to France and left England itself vulnerable to invasion. And the fallout from John's refusal to abide by the provisions of Magna Carta (1215) sparked the rebellion of the barons, who invited the French heir apparent Prince Louis to invade England. In the Battle of Sandwich of 1217, two moonlighting admirals clashed in adapted fleets. In this, the first naval battle of any substance, Eustace the Monk, a holy man and pirate, led a French invasion fleet against a rival fleet commanded by Hubert de Burgh, chief minister of England.

De Burgh's ships were merchantmen offered up by the Cinque Ports. Since the eleventh century, this ancient confederation of port towns in south-eastern England had supplied the Crown with ready-crewed ships for fifteen days each year in return for exemption from certain taxes and other privileges. Originally consisting of Hastings, Romney, Hythe, Dover and Sandwich, over time other port towns in the area became 'limbs' of these founder members, helping them to meet the Crown's maritime needs. For, apart from a few ships, the Crown maintained no fleet of its own; rather, it relied on requisitioning merchant and fishing vessels. Just as the Bible speaks of ploughshares being hammered into swords and back again, so these ships would be adapted for war – the high-waisted cogs being particularly useful in raining down missiles on lower enemy ships – and, in peacetime, converted back into merchantmen.*

In the century or so following Sandwich there were few sea-battles of any note. The main aggressors in the Channel were pirates, often operating with the tacit or express permission of the French or English governments, preying on the flocks of mercantile vessels wobbling along the coasts between ports. Until the Hundred Years War, in which the territorial ambitions of England and France eventually boiled over into a long-running, stop-start conflict – specifically ignited by Edward III's claim to the French throne – formal naval battles between nations in the Channel were comparatively rare. Coastal galley-raids were more common. And when they did happen, there was nothing especially different about the style of the fighting from land battles. Usually preceded by a hail of arrows from archers on the fore or sterncastles, enemy ships would draw alongside and grapple inextricably together. At Sandwich, the English sailors flung quicklime in the faces of the enemy for good measure. Parties of armoured knights

---

* The exception was the galley. By far the most effective instruments of naval warfare, these low, slender vessels could mount devastating raids upon stretches of enemy coastline. Oar-driven, the galleys were fast, manoeuvrable and could easily come up rivers to attack targets inland. Wedded to its policy of requisitioning merchantmen, England did at various points (notably in the reign of King John) have some of these specialised warships. But France and its continental neighbours possessed fleets of them and devastated the Cinque Ports accordingly.

would then board their rivals' ships and engage, battlefield-style, on the swaying decks at close quarters with swords, maces and spears. It was almost as though the sea was an incidental presence.

It must have seemed this way too, on 29 August 1350, when, sharply attired in a black velvet jerkin and beaverskin cap, King Edward III stood buoyantly in the bows of the *Cog Thomas*. For three days his fleet had ridden at anchor in the Channel between Dover and Calais, waiting to intercept a Castilian fleet leaving Sluys for the Spanish mainland.

Although a recent truce had suspended long-running hostilities between England and France (later known as the Hundred Years War), the Flanders-based Castilians – originally commissioned by the French to prey on English coastal ports – continued to plunder English shipping. Most recently, Charles de la Cerda, the Castilian captain, had captured English ships carrying wine from Bordeaux and slaughtered their crews. Not only were these individual tragedies affronts to the King: such piracy created insecurity in the Channel, endangering English commerce and the revenues on which the Crown depended. Reliable sources in Flanders had supplied details of the route and departure time of the Castilian voyage; Edward vowed revenge.

To eyewitnesses, that afternoon the King was at his sprightliest, calling for the ship's minstrels to play a fashionable German tune, compelling Sir John Chandos, architect of the recent victories of Crécy and Poitiers, to dance for him. Gaiety (perhaps enforced) ruled the deck. Then a lookout in the top sung out a sighting of one Spanish vessel, large and heavily armed, then another, then another. The minstrels abruptly downed their instruments. The King and his knights each downed a glass of wine (most likely Bordeaux) and donned their battle-helmets. The ships huddled closely into battle formation. From their decks came rousing trumpet-blasts; the Battle of Winchelsea was in the offing.

As the Castilian vessels bore down fast upon those of the English, the King commanded: 'Steer at that ship straight ahead of us. I want to have a joust at it.'[4] The two ships rammed one another with such force that the King's *Cog Thomas* sundered along its seams, badly shipping water. The knights scurried below and bailed for their lives. Eventually the grapnels were flung, the rival ships were yoked together and the melee began.

By this time, the interlocked ships were off the Winchelsea coast, the action clearly visible from the shore. As dusk fell, spectators lining the cliffs watched a battle between adversaries fairly evenly matched. The Castilian vessels were higher than those of the English and well armed, making the boarding of them difficult. The King's *Cog Thomas* and the Prince of Wales's vessel *Bylbawe* were so badly damaged that they sank (their occupants escaping to other ships). In ships and knights, there were heavy casualties on both sides.

Victory of a sort went, in the end, to Edward and his men. But like many of the conflicts of the Hundred Years War, it was ultimately inconclusive. They captured or sunk fourteen of the Castilian vessels, while the rest fled for the safety of French and Flemish ports; these would soon again be harrying English shipping. As the sun sank over the corpse- and weapon-strewn Channel,* the surviving ships of Edward's fleet limped into the ports of Rye and New Winchelsea, their trumpeters sounding wearily triumphant calls.

From the Lookout, no ships are now to be seen. Fields, not waves, roll away to the eye's limits, checked distantly by Dungeness. But in the Middle Ages, the Winchelsea townsfolk would have enjoyed a very different view. The sea lapped at the town, blanketing the present landscape of fields and marshes. Ships unloaded at a harbour downhill from the medieval Strand gate, ruinous now but still bestriding the road as a portal. Behind us, under Lookout Cottage, lurks an early fourteenth-century

---

* Incidentally, in the mid-nineteenth century, a fisherman found another medieval ship's trumpet on the shore at Romney, a few miles up the coast from Winchelsea. Had it been lost overboard during one of these strange naval battles?

cellar for the storage of Gascon wine taken up from the quay. Now, however, the scenery is sweetly rural. As from Billingsgate, the sea has absented itself from Winchelsea.

Perhaps the finest of the medieval seals depicting ships' trumpeters is that of Winchelsea (*c*.1300) and that is what has drawn me here. The town's seafaring fortunes reached an early crescendo in the Middle Ages, and then met an early end. Medieval Rye and Winchelsea stood with Camber at the points of a large, triangular estuary, where the rivers Brede, Rother and Tillingham debouched into the sea. By 1191, Rye and Old Winchelsea were sizable ports, home to enough vessels for them to become 'limbs' of the Cinque Port of Hastings. Later they would be raised to the status of full members of the confederation.

Old Winchelsea had originally, and unwisely, been built on a shingle spit that extended offshore from modern-day Camber. In the thirteenth century, a series of savage storms tore away the shingle and then destroyed the town. But such was the place's importance to the wine trade that it was completely rebuilt nearby on higher ground. In 1328, this small port held its own among the larger port cities in England, being ranked ninth in the country for wine imports. The Mayor of London was consulted on the building of New Winchelsea, the present town, and Edward I himself took a close interest in its layout, a grid pattern reminiscent of continental settlements (and a very early example of an English planned town). This echoed the entwinement of Old/New Winchelsea's citizens with France, who shuttled with ease between the ports of Normandy and Gascony, at least until the tense conflicts of the Hundred Years War.

But today, as if to underline the medieval ambiguity between land and sea, both Rye and Winchelsea are more or less landlocked. In the early modern period, the large estuary that made them sea-towns was gradually silted up, stoppering their harbours; their rivers, the Brede and the Rother, were rendered unnavigable and shrank to mere streams. Though small fishing boats may still get out of Rye, the fleet is a fraction of what it was. And now, it is surreal to walk around Winchelsea's thirteenth-century grid of streets and think of its medieval import. Little above ground testifies to the fact that it was once a harbour. Grassy verges and pretty whitewashed cottages evoke, in the view of one authority,[5] not so much an ancient city as a garden suburb. Yet its medieval shipscape lies shallow in the ground, and it is this that I have come to seek.

The church of St Thomas tells of Winchelsea's early patronage of and subsequent forsaking by royals. It was once nearly cathedral-sized but is now only a third as big. Today, a crablike building squats in the square churchyard, victim of warring and weathering, a grand chancel* left with the clawlike remains of its transepts. My friend Michael is excited to learn that Spike Milligan lies in the churchyard. Jo, curator of Winchelsea museum, has kindly agreed to show me a portrait of a medieval ship.

Inside, at a pillar in the north-east corner, Jo peers for a moment then points to an area of masonry near its base. I squat and squint in the low light, seeing only a mass of scratches and dents, until Jo traces the faint outline of a medieval cog, scribed inexpertly into the stone. From the grey surface it emerges as though from mist: spidery stays and rigging, fore- and sterncastles, mast, hull. It's a marvellous survival, overlooked for centuries until spotted by a keen church-crawler. And, subsequently, other ship-effigies like this one were found in the medieval churches of the Cinque Ports. Their purpose – prayers, memorials, muster-points – is still conjectured. But their elusiveness emphasises the distances of these ships from us now.

However, Winchelsea's medieval shipscape is not all so elusive. Under the town, below some thirty-three buildings from later centuries, there survives a remarkable honeycomb of vaulted cellars, built in the Middle Ages for storing the barrels of wine on which the town depended. No other place in

---

* The part reserved for the clergy at the east end of the church, where the altar is located, as opposed to the nave at the west end, where the congregation sits.

Britain has as many surviving examples of these small, mysterious chambers.

After a pause by Milligan's grave (Michael, a Limerick man, is much taken with its Celticism), we leave the churchyard and turn several corners to a house which once would have harboured a sea view. It's owned by John, a retired surgeon, who bought it years ago and found a long-lost cellar underneath. I like the way Jo and John treat Winchelsea's monuments with an easy familiarity. They talk idly of the welfare of the Strand Gate (hit by a vehicle last week, apparently), and enthusiastically of the newly found cellars emerging under renovated houses.

Shooting us a keen glance, John asks Michael and I if we know anything about these cellars, or medieval masonry; we must admit that we do not (of Gothic, I've forgotten more than I know). Nonplussed, he leads us down a flight of stone steps through a large hatch in front of his house, like the route of barrels to beer cellars; down these stairs, wine-barrels were once lugged into dark security. Actually, John says, the cellar was not used solely for storage. This becomes obvious when he lights a string of naked bulbs to better illuminate the cellar's fine masonry vaulting. A series of rib-vaults bestride us, their stonework beautifully tooled even after centuries of being backfilled with rubbish. Here, he says, customers could sample newly landed Bordeaux in surroundings far beyond those of an ordinary storehouse.

These Winchelsea cellars were the salehouses and strongrooms of the town's wine merchants. That there are so many illustrates its medieval prosperity and the extent of its sea-trade at that time. But they are more than just architectural curiosities; they are vessels, of a kind, long moored underground. Winchelsea's cellars are stone counterparts to the lost ships' holds in which the wine was carried, while their pointed ribs resemble the lost prows which cleaved the Channel. We could take them as effigies for the many cogs that once clogged the harbour here. And in one cellar under Blackfriars Barn, this monumentalising quality goes further still.

One of the main sources for the Battle of Winchelsea is Jean Froissart, a contemporary French courtier and chronicler who based his account upon the testimony of aristocratic eye-witnesses. As a result, we hear much of the deeds of Edward and his retinue but little of the ordinary crews. But in 2015, in the cellar beneath Blackfriars Barn, an unkempt former civic building, something like their testimony was found. For centuries, the space had been used as the town dump, gradually filled with rubbish and then forgotten. A recent clearance uncovered the north wall of the cellar for the first time in centuries. Incised in the plaster is a bewitching seascape: an impressionistic melee of ships and rigging, provisionally dated by experts to the fourteenth century. While the plaster was wet it was quickly graffitied, the ships' lines most probably drawn with fingernails.

It's thought that this Winchelsea seascape represents a folk memory of the Battle of Winchelsea, drawn in the mid-1300s at around the same time as the battle occurred, quite possibly by returning sailors for whom it figured as a glorious moment in their lives and that of the town. Down here, in the reek of the cellar heavy with grape-scent and must, they left a rendering of their battle memories, of the knots of ships in chaos upon the Channel. With the siltation of Winchelsea's harbour in the early sixteenth century, such memories of ships passed out of mind. With the uncovering of this plaster seascape, the impressions of an ordinary sailor seem to live again.

Not that long after the Battle of Winchelsea, in the last century of the Middle Ages, ships began to specialise. Shortly before the English victory at Agincourt in 1415, Henry V realised that he needed a more reliable navy to quash French sea power and guarantee any victories on land. So he ordered the building of four colossal new ships of war. Two of these, the *Trinity Royal* and the *Holigost*, were existing ships that were much enlarged, while the other two were bespoke, the 1,000-ton *Jesus*, built at Smallhythe, and the even vaster *Grace Dieu*, built at Southampton. They were the aircraft carriers of their age, designed to be warships, and warships only.

Of them all, the *Grace Dieu*, Henry's flagship, was the largest and most impressive. But, as if to illustrate England's continuing adolescence in maritime affairs, she made only one, failed, voyage: a short Channel cruise that ended in mutiny. Despite her absurd career, she marks a watershed moment, when bespoke warships began to be built for England's Kings, the most covetously naval of whom would be Henry VIII (*Mary Rose*, his pride and joy, would be as equally doomed). Subsequently, *Grace Dieu* mouldered for a while at anchor, before being permanently dry-docked upstream on the river Hamble. In 1439, a bolt of lightning struck her mainmast and, blazing ferociously, she collapsed. Today, the lowest tides sometimes show her bones.

It was in mud docks, mere riverbank excavations, that *Grace Dieu* and her sisters took shape. When ready for launching, they were freed by breaking through the earth wall that separated them from the water. It was as though the gap between the man-made and the natural was still ambiguous; it was as though the natural could easily reclaim the nautical, if it so chose. Appositely, now, medieval vessels are only to be found in the mud, whence they are excavated as though in parodies of those launches.

Sometimes, they may only be discerned by the imprints in the earth that they left behind, their timbers having long since perished. Similarly, some aspects of medieval seafaring require us to observe the contours and customs of later ages to deduce how it was done. We don't, for instance, know exactly how medieval ship's trumpeters operated. But we do have clearer accounts of how their Tudor successors did. In 1582, for instance, the trumpeter aboard Sir William Monson's flagship was required to:

> have a silver trumpet, and himself and his noise to have banners of silk of the admiral's colours. His place is to keep the poop, to attend the general's going ashore and coming aboard, and all other strangers or boats, and to sound as an entertainment to them, as

also when they hail a ship, or when they charge, board or enter her . . .[6]

Climbing out of the cellar, we thank John for his time and walk back to the churchyard. On the way, Jo speaks of how her late husband once served as its Mayor, in an unbroken line of succession stretching back to its foundation in 1288 (including Gervase Alard, the first English Admiral; he is buried in the church), and how she herself had recently been made a Freeman. Like the City of London, Winchelsea is governed by an ancient Corporation, now largely honorific but still a potent cultural force. We part under the town sign, bearing the arms of the Cinque Ports (a lion's forequarters joined to a ship's stern), atop which rides a golden depiction of the ship upon Winchelsea's ancient seal; two ships' trumpeters glint against the sky, poised to sound a hailing, a charging or a boarding.

Romans believed that Neptune once punished a miscreant people by flooding the world. When it had been totally immersed, he and the other gods paused their persecution when only two survived: a man, Deucalion, and a woman, Pyrrha. Then, Neptune's merman Triton raised his spiral horn and, with something not unlike a trumpet-blast, bade the waters retreat. It was the Roman writer Ovid who, in his *Metamorphoses*, sketched the most compelling retelling of this reverse-flood: 'the sea recovered its shores / the rivers, though full, were confined to their channels; the flooding receded; the hills were seen to emerge. The earth rose up; as the waves died down, dry land expanded.'[7]

This myth could be an allegory of what happened here, in Winchelsea, when silt blocked the harbour and land flooded the town's environs. And it makes me wonder whether the Billingsgate Trumpet, if blown, might have the reverse effect, not only on this town but also on other stoppered medieval ports – resurrecting their admirals, raising their wrecks, refloating their fleets, conjuring back into being a lost shipscape, ancient and granular, in the force of one, long, gilt-edged blow.

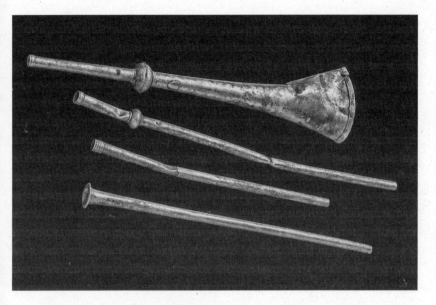

4. The thirteenth-century Billingsgate trumpet, found in 1984 in medieval river silts next to Billingsgate

Trophy

The Golden Hind *Chair*

PLYMOUTH, DEPTFORD, LONDON AND OXFORD
ELIZABETHAN

Held aloft, the face still expressed fleeting shock at the axefall. As blood pattered from the ragged neck, the ruddy cheeks kept their colour momentarily, then blanched. Below, the headless corpse slid off the block and slouched sideways. Holding up the head of his colleague, the captain looked out over his company, assembled dolefully in the bay, and boomed 'Lo! This is the end of traitors!'[1]

In the background, their four ships rode at anchor in a foreign sea. Englishmen had never come this far south, or still yet traversed the hellish strait awaiting them. Only one ships' company had successfully navigated it before – and those Portuguese had been decimated by starvation and disease on the way home. They had also mutinied.

To Francis Drake, the gentleman adventurer Thomas Doughty had seemed a promising recruit. Soldiering in Ireland, Doughty had shown himself a man of action; Drake also liked his aristocratic connections – and perhaps most importantly, the money he was prepared to invest in this voyage. But, for reasons still unclear, Doughty decided to spend the voyage sowing discontent among Drake's crew, attempting to secure the captaincy for himself. This had been going on even before they had crossed the equator, and by the time they reached the Argentine coast Drake could stand no more. To avert mutiny amongst his men, he tried Doughty (with dubious legality), condemned him to death, and urged his men to press on with their passage.

Part of the problem with this voyage was the division between mariners and gentlemen, a distinction we might trace back to medieval seafaring, with its awkward grafting of landsmen's ways onto mariners' methods. Gentlemen were then still just passengers aboard ship until battle was joined, when they fought as if they were on land; centuries later, Doughty's gentlemen were the same, lording it over the sailors and recoiling from seamanship. The disparity threatened to spoil Drake's unprecedented voyage out of the Atlantic and into the unknown, where all hands were truly needed on deck.

As the ships' company prepared the stores and the ships for the coming passage through the Strait, there were mutterings, dissension, embers of ill-feeling. So, nine days after Doughty's execution, Drake drew the mariners and gentlemen together and addressed them again: 'Here is such controversy between the sailors and the gentlemen, and such stomaching between the gentlemen and the sailors, that it doth even make me mad to hear it. But, my masters, I must have it left, for I must have the gentleman to haul and draw with the mariner, and the mariner with the gentleman.'[2]

Apart from the virile economy of language, so characteristic of the Elizabethans; apart from the staggering social implications – gentlemen to dirty their hands?! – Drake's words were epochal. They were the first crack in the feudalism that still prevailed aboard ships and was hindering their oceangoing promise in this new age. They were recognition that oceangoing required a break with the land in every sense. Ships still functioned hierarchically – but, now, all aboard were levelled to the requirements of the ocean. It was with these famous words, uttered in a place unknown to medieval mariners, that English seafaring began to leave its juvenile phase. In this stellar time, England learned to cross the oceans and could not resist the temptation to bring home trophies of the new worlds encountered on the other side: gold, ore, slaves, hostages, land, dreams.

Around the quay, the sunlight pools upon the frigid paving, twinkling the veins of quartz in the granite. It's a bright, fiercely blue morning, cold as only February can be, a cold that no layers keep out. We stalk through a city deserted but for gulls shrieking savagely from bins and gutters. Immortal pests.

Trophy

Yesterday's revelry has deprived us of our humours. It had been one of those golden afternoons trundling westwards in a railway carriage, where there were no calls upon our time, where good beer flowed endlessly from the buffet car, where the excitement of going west rang in the heart. Now we pay.

We are pale wraiths who need remedy – coffee at least, though we have done nothing to deserve it, nor the bacon sandwiches we are coveting. We argue the toss about ketchup and brown sauce. My brother favours the former, I the latter, but we fling this trifling distinction to the wind. Down along the craggy walkway over the sound we inspect the debris left by the storm of the last few days. Then back up to Sutton Harbour, where Plymouth's old core has been engulfed by later urban sprawl. We scan the shopfronts with bleak dismay. Nothing is open except the door of a hunched stone building that looks much older than its neighbours.

To pass the time we go in. This ancient building turns out to house a newish second-hand bookshop. As we enter, ducking under the low lintel over the threshold, a man behind the counter stands up. He's not your archetypal second-hand bookseller. No, he looks like an old tar, stout, shaven-headed and one-eyed. Apropos of nothing he waves around, indicating the shop, and growls 'this used to be the Custom House in *Drake's* time. For fiction go upstairs.' Then he slumps back down to his desk, on which a small notebook lies open like a ledger.

Drake. The word lies on the table between us like a gleaming coin. It calls up images of three-masted ships printed upon westerly horizons or jostling in the harbour we have just skirted; hangovers make me far more fable-prone than usual. Images of ruffed adventurers, the first Englishmen to feel the tropics underfoot, whose names are perhaps more admiringly uttered than their deeds are exactly remembered: ornaments in a nation's firmament. I take stock of the cramped interior of the bookshop, picturing lordly sea-captains counting coins. My brother fidgets with hunger and boredom.

That encounter was nearly a decade ago, but I still remember that old bookseller and the way he had uttered the

65

name 'Drake' as though the captain were still with us, listening. Plymouth has long lain under the spell of the Elizabethans, ever since their globe-trotting and Armada-beating deeds. Of all England's port cities, I find it stands alone in still carrying the atmosphere of Elizabethan seafaring, that great coming-of-age when England's horizons expanded westwards and new, troubling yet staggeringly lucrative cargoes began to be carried to and fro over the Atlantic; an age when a recognisably naval style of warfare began to emerge, hinging upon the gunfounding and gunnery in which the English were then unequalled; but an age in which, paradoxically, England owed much of its maritime fame to the teachings of its greatest foe.

'Let it be the thickness of half a finger at the least, for the weightier that it shall be, so much shall it be steadier to make the altitude . . .'[3] These were the preliminary instructions for making an astrolabe, carried to England by the navigator Stephen Borough in a bundle of Spanish papers. In 1558, he had journeyed to Seville to learn Spanish navigational arts – at that time, with the Portuguese, foremost in the world – in exchange for his own expertise in seafaring around the near Arctic. A few years previously, in 1553, Borough had been the master of the *Edward Bonaventure*, the first English ship to navigate to the White Sea and open trade with Russia.

In Seville, Borough was received with great respect by the Spanish and presented with a pair of perfumed gloves, their symbol of the qualifications and office of an examined pilot. In return, he told them all he knew of Arctic navigation. It was an extraordinary cultural exchange, made possible because England was still under the rule of Queen Mary I and her husband Philip II of Spain. And it was to have momentous consequences for the future of the two countries. For Borough returned to England with a copy of the renowned Martin Cortes's*

* Not to be confused with the son and heir of conquistador Hernan Cortes.

navigational training manual, which would subsequently be translated into English as *The Arte of Navigation*.

The Arte was the first navigational handbook in English; moreover, it was a guide to deep sea navigating, far out of sight of land. Even as late as the 1550s, English pilots remained nervous about straying too far from the coasts, relying on the daymarks and rutters that Chaucer's Shipman would have known. England's first transatlantic voyage, in 1497, had been led from Bristol by Zuan Caboto, a Venetian (anglicised as John Cabot). But after *The Arte* appeared, everything changed. For not only did it contain instructions for oceangoing navigation and cosmography, it also explained how to make and exploit navigational instruments. Among these were the lodestone, the cross-staff – for taking the altitude of the Pole Star – and the astrolabe, the crucial instrument by which a mariner could determine a ship's latitude from the sun's altitude at midday.

'To take the altitude of the Sunne, hang up the Astrolabe by the ring . . . Then look upon the line of confidence . . .'[4] Circular, of brass and roughly a handspan's breadth, the astrolabe was a beautiful, occult-looking object, trophy-like in how it was coveted and in the promise that it held. Devised by Portuguese navigators from a much more ancient design, within a few years of *The Arte*'s publication every aspiring English mariner possessed one. With these instruments, deep-water seafaring become far more frequent and egalitarian, no longer the preserve of a select few with suicidal bravery or esoteric knowledge.

Borough well knew the revolutionary impact *The Arte* would have. On his return to England, he pressed for its translation and publication, ostensibly out of benevolent motives; of him the translator Richard Eden stated in his preface: 'he desireth the same for the common profite to be common to al men'.[5] But unbeknownst to him, Queen Mary was dying. On 17 November 1558, Queen Elizabeth I ascended the throne. Three years later, in 1561, *The Arte of Navigation* was published by the Queen's Printer, Richard Jugge. Borough had plucked this diplomatic fruit from Sevillian orange groves now off-limits to

Englishmen. And it would shortly be put to most undiplomatic uses by Elizabethan mariners.

We stand, Dad and I, looking at the Elizabethan houses tottering down Looe Street. Dad hails from Looe, but we are here in Drake's wake, looking at the site of his long-demolished Plymouth house. Outwardly, there's not much except the crabbed and gabled frontages to identify this place with him. Further down, there is a pub wedged between the houses. We look at our watches, then at one another.

We bow through the low entranceway. Inside, a taproom of Elizabethan proportions, long and narrow, about as long as the narrow frontage of the pub is tall. Though walled, floored and ceiled in much later timbering, the shape of this room rings true. Allegedly Drake himself would have looked in here – not impossible, though impossible to prove. Any ships' timbers in here, I ask the landlady. She nods at the stair and says it's built around the mast from a Spanish galleon. I salute it with my pint.

When Elizabeth I came to the throne, the medieval trade in wool and cloth with Northern Europe that for centuries had sustained England's economy was faltering, though not entirely in the doldrums. Bickering with Spain was steadily escalating, the result of its presence in the Netherlands disrupting the wool trade and its scheming to place a Catholic ruler upon the English throne. As a result, England could no longer wholly rely on the Old World for its prosperity and its merchants began to look for new markets. But compared with the mammoth empires of the Continent it was a fairly insignificant and underpowered island. The break with Rome had given it a new footing in the world, but Henry VIII had spent profligately and almost bankrupted the nation. So, under Elizabeth, state-sponsored explorer-pirates set out westwards, looking for new lands and new streams of revenue, fair or foul.

Fair or foul. In 1562, John Hawkins discovered the Middle Passage. Ever since the fifteenth century, the Hawkinses had been embarking on the trading voyages that had established them as one of Plymouth's wealthiest families. William Hawkins had traded not only with the Continent, but also with Guinea (West Africa) and Brazil. Shortly after Elizabeth's accession, his son John made several voyages to the Canary Islands, learning there of the lucrative possibilities of trading enslaved Africans between the West African coast and the Spanish West Indies. After securing investment from a syndicate of well-heeled Londoners, including Sir William Winter,* surveyor of the navy, Hawkins sailed from Plymouth in October 1562, arriving at the Guinea coast at the year's turning.

He abducted at least 300 African people and transported them across the Atlantic to Hispaniola, in that second and most notorious leg of the 'triangle trade'. The Middle Passage was a route to riches, though not of the fabled kind to lands afar as would be sought by other explorers. This was an evil passage over known seas.

Arriving in the Caribbean, Hawkins crept between small, out-of-the-way ports, offloading his human cargo in exchange for gold, ginger, sugar, pearls and hides. He avoided the Spanish administrative centre, San Domingo, because he was trespassing on Spanish and Portuguese monopolies, which stemmed from a papal decree of 1493; indeed, parts of the precious cargoes for which he had traded the enslaved Africans were impounded from ships he had sent to Seville, possibly in an attempt to secure a Spanish licence for his activities. Despite the seizures, the voyage was immensely profitable. Hawkins was the first Englishman to commodify a people, scribing an inhumane trade in them between Africa, the Americas and England.

Inhumane to us, of course, but not to the Elizabethans, or Elizabeth herself. For his next two voyages, of 1564–5 and 1567–9, Hawkins chartered from the Queen an aged carrack, the *Jesus of Lübeck*, and was allowed to sail under the royal standard. These voyages yielded even greater profits, but the third would be the last. It ended with a disastrous fight with the

---

* Later a prominent landowner in Lydney, where his body (although not his 'bowelles') was buried.

Spanish in a Mexican port, with only two of Hawkins's eight vessels making it home – one of them, the *Judith,* commanded by his young kinsman Francis Drake.

For his trouble, the Queen granted Hawkins a coat of arms. His new crest sported a 'demi-Moor proper bound in a cord'. It spoke nakedly of the astonishing profits of slavery – a 60 per cent return on investment for his last voyage – and of the subsequent social elevation that this bought him. And it spoke of the founding hypocrisy of the slave trade which would reign in Britain until abolition. For the trade went on over the horizon; proportionally few of the enslaved people appeared on English quaysides. Hawkins could bear his crest with an easy mind. With the (English) trade in its infancy and most of the population unfamiliar with its horrors, his 'demi-Moor' was simply a trophy to parade.

Draining our glasses, we leave the hold-like taproom and walk into a fresh wind on Looe Street. Talk of Drake means a visit to the bookshop that had so transfixed me all those years ago. But the dust on the windows shows that, alas, it has long since closed, so instead we walk to the harbourside where Hawkins' ships would have come in, though the quaysides they touched have long since been refaced.

From here, Drake led his own expeditions to the Spanish Main. Having learned lessons in seamanship and deportment from Hawkins – who conducted himself like visiting royalty in the slave-markets, much impressing the Spanish colonists – Drake spent the early 1570s raiding Spanish settlements in the Caribbean. In 1573, he seized a most desirable prize: the Spanish plate train, carrying silver from Peruvian mines across the Panamanian isthmus to the Atlantic coast, whence it was loaded onto ships and ferried across to the homeland's coffers. Drake's share was £20,000, enriching him enough to become a Plymouth grandee. And he brought back something arguably more valuable than mere loot: a glimpse of the Pacific from an isthmian palm tree.

*

While Drake dreamt of that unknown ocean, a very different expedition was being fitted out on the other side of England. In 1575, the Russia Company – for which Stephen Borough had voyaged – issued a patent to Martin Frobisher and Michael Lok for a voyage to find the north-west passage above North America to Cathay, to assess and exploit the potential for trade. Frobisher was a Yorkshire pirate, often found wanting in seamanship but not in aggression; Lok was a merchant adventurer and the Russia Company's agent who put up a lot of the money for the scheme. In the middle 1570s, tensions between Spain and England had eased, and a more amicable relationship seemed possible. The north-west passage would avoid Spanish colonies and conflict therewith, a less provocative route to riches.

Frobisher was to command the *Gabriel*, the largest vessel in a modest flotilla of three. In June 1576 they left the Thames, promising to give Lok on his return 'the first thynge that he found in the newland'.[6] Touching at Greenland, then Baffin Island, Frobisher arrived in the sound that he would soon name after himself. As they pushed northwestwards, Frobisher and his men met Inuit, the first Englishmen to do so, initially with friendly contact. But on 20 August the Inuit apparently abducted five of the English; in reprisal, Frobisher abducted an Inuk. And from a little island in this tract of wintry sea, one of his men picked up a black stone.

Shortly afterwards they returned to England, deterred by poor weather from further exploration. Back in London, Frobisher implied that the passage likely lay further northwestwards up Frobisher Strait, and that he would have found it were it not for the onset of the hurricane season. And he exhibited his captured Inuk to public marvel: a 'strange Infidel, whose like was never seen, red, nor harde of before, and whose language was neyther knowne nor understoode'.[7] His name has not survived.

But it was the black stone which most interested Lok. When he weighed it in his hand, it seemed to glint favourably. In his account to the Queen, Lok explains how he gave a piece to 'Mr Williams, saymaster of the Tower', who found nothing; he then gave a piece to 'one Whelar, goldfyner', who found nothing; he sent it around London's assayers until he found one who confirmed his gold-lust. In January 1577, John Baptista Agnello showed Lok 'a very little powder of gold, saying it came therout'. When quizzed by Lok as to how he had been able to extract gold from the stone where others had failed, Agnello replied that one must learn to flatter nature.[8]

The search for the north-west passage was not only about temporarily avoiding conflict with the Spanish: it was a quest for England's own source of New World wealth in emulation of Spain, which had grown obscenely rich on the metals of the Americas. And it was this quest which seems to have sidelined the finding of the north-west passage in Frobisher's second and third voyages. After all, if the black ore happened to contain gold, then the enriching ambition of the north-west passage might be fulfilled.

Immediately plans were made for a new expedition, this time to mine more of the black ore so that it could be assayed further and exploited for the gold allegedly found by Agnello. Leaving Harwich in May 1577, Frobisher in the *Gabriel* led another flotilla of three around the northernmost tip of Scotland, touching at Greenland as before, and found more ore on the northern shore of Frobisher Bay. Of this they mined 160 tons and returned to England, taking with them three more Inuit for public exhibition, to keep alive interest in the project. Their names were Kalicho, Arnaq and Nutaaq.

Meanwhile, in Plymouth, Francis Drake was preparing another expedition. Relations with Spain had soured again, the result of yet more sabre-rattling in the Netherlands. Accordingly, the Queen assented to a new proposal from Drake so that she might 'gladly [be] revenged on the King of Spain for diverse injuries'.[9] The ostensible mission of the five ships fitting out at Plymouth was to trade with Alexandria. In fact, they were to voyage to the South Seas and there plunder Spain's possessions along the west coast of America, in the Pacific, where King Philip expected to meet no foe. In his flagship *Pelican*, Drake led his flotilla out of Plymouth on 13 December 1577. They would be gone for three years.

Meanwhile, assays of Frobisher's second, larger shipment of ore were inconclusive. Far more would be needed to winnow out the gold. In 1578, Frobisher led a third voyage of fifteen ships off to Baffin Island. Over the Arctic summer Frobisher's men mined over 1,000 tons of the ore. But the voyage had another aspect too. On this, as on the previous journey, Frobisher was accompanied by Edward Fenton, a talented navigator who was to be governor of a proposed English colony of 100 men on Baffin Island during the winter of 1578–9. Unfortunately, crucial parts of their prefabricated blockhouse were lost at sea; what would have been the first English settlement outside Europe was written off. Before they left, however, Fenton oversaw the building of a sacrificial stone tower, erected to see what the region's winters would do to a building. 'Fenton's watch-tower' was the first English stone-built structure on the North American continent.[10]

In September 1578, Frobisher returned to England with ore-laden holds. Creditors of the first three voyages demanded returns; the assayers subjected the ore to every conceivable test in a purpose-built furnace in Dartford; it transpired Frobisher had laboriously shipped across the Atlantic a valueless form of fool's gold. The venture collapsed amidst much recrimination, damaged reputations and Lok's ruin – as late as 1615 he was still being sued for these debts.

As Frobisher's star waned, others waxed. On the far side of the world, Drake was navigating the Straits of Magellan, soon to come upon the easy, silver-laden prey of the Pacific. And another adventurer, Humphrey Gilbert would bring back to the Queen, not worthless ore, but the possession of a new land. A few months before Frobisher made his final return from the northwest, in June 1578, Gilbert received an extraordinary royal patent that entitled him to search out 'remote heathen and barbarous landes' over which he could enjoy near-monarchical possession.

Like Hawkins and Drake, Gilbert was a Devonian, but he was a little more unhinged – a soldier-philosopher prone to bizarre visions of dragons and angels. Nearly a decade before Frobisher's first voyage northwest he had been petitioning for just such an expedition; having failed, he went to Ireland as a captain to quash rebellions there, forcing the vanquished to process towards him along a lane flanked with the impaled heads of their families. After returning to England he advocated for the better education of men such as himself, the younger sons of gentry, in science and warfare in order to further the national interest. After receiving his patent of 1578, he made abortive colonising voyages.

Then, in June 1583, he left Plymouth in his flagship *Squirrel* with a fleet of five ships. In August they reached Newfoundland, where Gilbert landed at St John's and, after a skirmish with the local fishing fleet, claimed the harbour and all land within 200 leagues' radius for Queen Elizabeth. It was the first English possession in the New World since Cabot's excursion of 1497, and was symbolically marked by the leaden arms of England staked into the soil upon a wooden pillar. However, Gilbert never lived to present this trophy to the Queen. Off the Azores, homeward-bound, *Squirrel* was lost in a storm, Gilbert calling out 'we are as near to heaven, by sea as by land'.[11]

A clutch of isles in the mid-Atlantic, the Azores made for a convenient stopping-place between the New and Old Worlds, and were the scene of frequent skirmishes with Spanish fleets. In 1591, the Devonian squire and sometime privateer, Sir Richard Grenville, sallied out alone in *Revenge* against twenty Spanish warships, fighting them for twenty-four hours until his ship was shot to pieces. More of a headstrong soldier than a shrewd sailor, Grenville, dying, ordered his master gunner to blow up *Revenge* in order to prevent her becoming a Spanish trophy. In the end, the gunner refused, Grenville died aboard the Spanish flagship and *Revenge*, taken as a prize, later sank off Terceira.

The last fight of the *Revenge* quickly became a celebrated episode, even though it was really a foolhardy action which squandered one of the Queen's prize new warships. And that same year, in a piece of trophy-prose, Sir Walter Raleigh immortalised his cousin Grenville's end: '[*Revenge* had] the masts all beaten over board, all her tackle cut asunder, her upper work altogether razed, and in effect she was with the water . . .'[12]

Of all the famous Elizabethan mariners, Raleigh was perhaps the most complex, being many things to many men – or one woman. Courtier, soldier, explorer, poet – here was yet another Devonian who quickly became a favourite of Queen Elizabeth; he was much younger than his contemporaries, though, and carried through into the Jacobean era the Elizabethan lust for trophy-hunting. As well as being Grenville's cousin, Raleigh was Humphrey Gilbert's half-brother; immediately after Gilbert's drowning, Raleigh financed an expedition to Virginia to assess it for colonisation. Though influenced by his half-brother, Raleigh did not literally follow in his footsteps. Others carried out Raleigh's expedition for him. His first major exploratory venture would be later, in 1595, when he sailed from Plymouth for El Dorado. During this voyage, following the brutal conquest of a Spanish colony on Trinidad, an elderly prisoner – who was in fact the island's former governor Antonio de Berrio – told of fantastic cities and wealth to be found at the rising of the Orinoco. Up that river they went, finding nothing, returning to England with empty holds; but the legend of the golden city became a kind of trophy to Raleigh, one he would prosecute and promote for the remainder of his life before his career was curtailed by axefall.

Soon after we go aboard *Golden Hind*, I bark my shin on a step – not a good beginning. Insistent drizzle, bracketed by two downpours, has soaked the entire ship. While the stone of the dock gleams under the wetting, the deck timbers just look dull and uncommunicative. Partially rigged masts loom overhead. The main deck is surprisingly small and is faced with mysterious carvings of knight's heads. Slippery little ladders lead in and out of the holds. Below deck the spaces are cramped and irregular. In the Great Cabin there is a long table with a captain's chair placed against the rear of the transom. Rain has soaked my trainers and socks; a chill steals up the small of my back. I find it hard to focus or be inspired by this wet hulk. As I pick between the cabins, I bang my head painfully on a beam. For heaven's sake.

So glittering were the achievements of the Elizabethan mariners that it can feel like the history of English seafaring begins here, that there was little before them but haphazard drift. And yet nothing remains of the ships in which the great Elizabethan voyages were made. It's difficult even to nail down their basic details, let alone pieces of their timberwork. Which is doubly strange, for this period is one fetishised in the hierarchy of England's maritime story, perhaps second only to the golden years of the Nelsonic navy. Yet we are almost entirely ignorant of the essential actors. We know so little about Frobisher's *Gabriel*, Gilbert's *Squirrel*, Grenville's *Revenge*, Raleigh's *Destiny*. We have few images of them and no detailed portraits of the vessels themselves; we have some textual descriptions, for sure, but from these arise no notion of the ships' quintessence, the quirks and foibles which ensouled them.

I'm aboard *Golden Hind* on a trophy-hunt of my own. It's a replica of the ship which began her life as *Pelican*, in which Drake circumnavigated the world. Launched in 1973 from the shipyard at Appledore, Devon, the replica has followed in Drake's wake, circumnavigating the globe; indeed, she has more miles under her belt than her namesake, having toured various ports across the world. Now she's permanently berthed at St Mary Overie Dock in the heart of London. And the replica is excellent for understanding the quiddity of these ships.

In December 1577, Drake left Plymouth in *Pelican* for the South Seas. Along the way they plundered many ships, the most important of which yielded a Portuguese pilot named Nuno da Silva, who knew the South American coasts and whom Drake gratefully commandeered as their navigator. Crossing the Atlantic on a south-westerly course, they spent sixty days with no land in sight, Drake navigating only with an astrolabe and da Silva's guidance. In June 1578, they arrived in Port St Julian, a little anchorage on the Argentine coast. It was here that Thomas Doughty was beheaded. Then they continued south to the entrance to the Strait of Magellan, named for the

Portuguese navigator who had traversed it before them. On this ominous threshold, Drake led a salute to the Queen and a short religious service, before renaming *Pelican* as *Golden Hind*.

Pelicans are not native to Plymouth, but they were a resonant bird in Elizabethan England. A belief that the mother drew blood from her breast to feed her young when food was wanting associated the pelican with motherhood, self-sacrifice and devotion. These attributes aligned with the image Elizabeth I wished to convey to her subjects; in 1573, the painter Nicholas Hilliard had depicted her in fine crimson trappings with a jewel-like pelican at her breast. By contrast, a hind formed part of the crest of Sir Christopher Hatton, foremost investor in the voyage, whose secretary Doughty had been. The renaming was a way for Drake to reaffirm his allegiance to this powerful Elizabethan magnate and to mitigate his beheading of Hatton's servant; the ship's two names also spoke of the way these Elizabethan voyages were double-acts between merchants and Crown.

After navigating the Straits of Magellan, Drake raided Spanish settlements and captured Spanish ships in the Pacific. Since they expected to meet no enemy, these were easy prey. The main prize was *Nuestra Señora de la Concepción*, a treasure ship said to be carrying over 350,000 pesos in silver and gold. After this, they 'careened' the *Hind* – pulling it out of water and rolling it onto the side to repair the keel – before navigating the 7,100 miles across the Pacific to the Spice Islands, negotiating a trade deal with the Emperor there, then navigating the 9,700 miles from Java to Sierra Leone without a map, to bring the *Hind* at last, laden with riches, into the English Channel. In a feat that would have been unthinkable even twenty years previously, Drake had become the first Englishman to circumnavigate the world. And as the *Golden Hind* finally touched the Plymouth dockside, he passed into legend.

It can be an abstract business, writing about ships one has never seen. The *Golden Hind* sounds apple-seed-small in the vastness of the South Atlantic or South Pacific. Yet, as this replica of her demonstrates, she was a huge presence at close quarters. I take a step back from her on the quay. Lodged here, in modern Southwark, she looks as alien as she would have done in the bays and bights of Asia. She's not as majestic as a warship, nor as businesslike as a merchantman, but is somewhere in between: a machine for carrying men and arms over the sea and for returning with trophies in that bulbous hold. Despite her then-new Elizabethan lines, there is a touch of the cog in her, perhaps the carrack too, like a newborn with her ancestors in her features (and perhaps those strange knight's heads were a medieval hangover, too).

Elizabethan shipwrightry saw the archaic cogs and lumbering carracks displaced by smaller, sleeker vessels. Warships and merchantmen became swifter and more heavily armed, their hulls perforated with cannon and gun ports. These had been pioneered under Henry VIII, most notoriously on the *Mary Rose*, in the sinking of which its low, negligently opened gunports had been a major factor. Subsequently, gundecks were repositioned to better marry them to the hull's centre of gravity. And while the stern remained high like a carrack's, the bows were dropped lower like a galley's, the better to mount forward-firing guns. Loosely known as 'galleons', this new breed of ships reflected the more manoeuvrable, aggressive character of Elizabethan seafaring.

It wasn't only the ships themselves that evolved. In the mid-sixteenth century Mathew Baker, master shipwright at Chatham and Deptford dockyards, became the first in England to commit a ship's lines to paper, advancing from the inexact methods of medieval shipbuilding and moving towards better communication in design and greater accuracy in construction through the use of mathematics to lay down the lines. His *Foresight*, launched in 1570, was the first of the new warships built along galleon lines, and was aptly named. Baker had designed and built Frobisher's *Gabriel*, and the *Revenge,* sunk off the Azores. Backed by the deft administration of John Hawkins, who had become treasurer of the navy in 1577, Baker's new galleons formed a key part of the English fleet which fought and outmanoeuvred the Armada.

We don't know who designed and built *Pelican*. Drake

5. Drawing by Mathew Baker of *c*.1580 showing a 'race built' galleon. A rare, and gloriously strange, contemporary depiction of an Elizabethan ship

probably commissioned a local shipwright with whom he had had prior dealings. She was funded partly by a state grant designed to stimulate the building of ships over 100 tons (regarded as the minimum size for naval service) and mostly by Spanish bullion. *Pelican* was 150 tons, with eighteen guns and a double-sheathed hull to keep out tropical shipworms. Most likely she was built at one of the shipyards at Coxside, on the east side of Sutton Harbour in Plymouth.*

If Drake's brief for her was ever written down, it may only have specified her outlines – masts, tons and arms – with the detail left to the shipwright; in any case, this has certainly been lost. And no detailed contemporary images of her exist. True, there are a few illustrations, but these are tiny, being incidental details on maps or portraits, and seem to be little more than generic renderings of a galleon, rather than showing any *Pelican/Hind*-specific details. Later illustrations are by necessity imaginative reconstructions. And textual accounts are tantalisingly vague. Nuno da Silva, for instance, describes how well she sailed but not how she looked, except to say, somewhat cryptically, that she was built to a 'French pattern'.[13]

When Drake returned to Plymouth in 1581, *Golden Hind* was ballasted with silver. The Queen and her fellow investors

---

* There is an unsubstantiated claim that she was built in Slaughden, Suffolk.

were staggeringly enriched by the profits of the voyage, which yielded them returns of 4,700 per cent. The Queen's share alone was enough to pay off that year's national debt and invest £40,000 in another venture.

In or around April 1581, *Golden Hind* set out for her last cruise. Compared with the voyaging of the previous three years it was unremarkable: a short beat up the Channel into the Thames Estuary. She was anchored at the Royal Dockyard at Deptford, where she was visited by the Queen. The ship was lavishly hung with banners, and banks of temporary seating were erected around her for the spectacle. A single wobbly plank bridged the gap between the dockside and the deck; once the Queen had walked over, it collapsed under the weight of a scurry of hangers-on. After formalities, the Queen, shrewdly, had Drake knighted on board by the French Ambassador (so as not to be seen to be rewarding a pirate in Spanish eyes).

Unprecedentedly, she then decreed that *Golden Hind* be preserved forever. No English vessel had ever merited full preservation, rather than cannibalisation, after her sailing life had ceased. Hurriedly a hull-shaped hole was excavated and into it the ship was positioned. Then she was left in the hands of the dockyard. Possibly she first rested on stocks to keep air circulating around her; more probably she was buried in mud up to her waterline. Originally, she was to be roofed over, but this was never carried out. By the end of that year Drake, the Queen and the Court had all moved on, leaving her earth-fast by the Thames.

Eighty years passed. Although she would have remained an object of fascination, little is known of her curious afterlife. But still people visited her, even caroused upon her – or at least this is what we might deduce from a sly reference in *Eastward Ho!*, a 1605 play by Ben Jonson. In it, the scoundrel knight Sir Petronel Flash suggests a boozy dinner aboard (who wouldn't?). But this may have been quite a risky business. In June 1618, the Venetian Ambassador saw her at Deptford, where she 'looked exactly like the bleached ribs and bare skull of a dead horse'.[14] By the time of the Restoration, she must have been so decayed that it would have been folly to board. In early 1662, the scene in the dockyard must have been a strange one. In the company of neat new frigates lay the wreck of this once-esteemed galleon, all masts askance, snapped spars, sagging decks. To the Deptford Storekeeper, in whose purview lay her fate, enough was enough. Consequently, in or just before 1662, he decided to break up the *Golden Hind*.

Will, a carpenter friend of mine, imagines what happened next. Having worked on the restoration of the *Cutty Sark*, he knows something about handling old ships. Even in a decayed state, demolishing the *Golden Hind* would have taken much time and labour. Most probably she was dismantled down to her freeboard, the remaining hull left buried and subsequently built over. Any sound timber remaining would have been carted away for reuse elsewhere in the dockyard. Will describes how the carpenters would have moved gingerly through the by now perilous ship, tapping each member for soundness.

After the carpenters had finished with her, nothing remained of the ship except the buried hull and a pile of sound timber in the yard.* According to Will, probably more survives of the *Hind* than we think, as the carpenters and joiners who dismantled her would almost certainly have pilfered the pile for little souvenirs – or trophies. I sympathise with this very human impulse, which may well have resulted in a great scattering of ship-shards around Britain: a new tool-handle, a whittled effigy, a handcrafted timber tankard. Such are the heirlooms of the *Hind* that might be lying unrecognised in our homes.

If you look to seaward by the Mayflower Steps, you see the way to the Americas framed by inclined cliffs but barred in the view by the black line of the Victorian breakwater. It seems to suggest that no enterprise leaves Plymouth these days, though of course this isn't the case. Sutton Harbour clatters with the angular poles of yachts tied up at the pontoons. It's more of a

---

* In 1975, archaeologists dug for the remains, with inconclusive results.

marina than a working harbour now, though a few wide-girthed trawlers loom over the pleasure craft. The harbour walls have mostly been refaced, but old patches of rubble stone can be seen here and there; the harbour and its tendrilled streets are famed for their cobbles, of which Plymouth apparently has more than any other English city.

Behind the harbour, the coastal pavement winds upwards under the lee of the great stone fort which Drake urged be built when he was Plymouth's mayor. It has been much enlarged and rebuilt since his time. To seaward, the headlands framing the sea slide further apart in the view; here, the great sound properly unfolds, plied by grey warships, the Victorian breakwater now like an enormous stone submarine, temporarily surfaced. Puffs of cloud tend to hang over it like gunsmoke. In 1588, of course, real gunsmoke hung in the Channel here, as the vast and creaking Armada of ships sent by the Spanish King processed up the Channel. Philip's patience with England had finally snapped – the endless plundering of his ships and American dominions called for decisive retaliation. The English fleet was punier, a fraction of the size of Philip's, but the Armada was gradually reduced by skirmish after skirmish as it creaked upchannel, scattered by fireships, then dispersed by storms. England had won a great victory – if the weather be counted amongst its arsenal.

Just beyond lies the bluff famous as Plymouth Hoe. It was from the harbour that Drake came and went, establishing his fortune and his repute as a navigator, but it was from the Hoe that he (legendarily) watched the Armada assembling in the Channel, although still finding time to finish his game of bowls. And as if to separate the man from the legend, there are no monuments to Drake in the harbour, but there is a fine statue of him on the Hoe.

The prominent bronze effigy of Drake that adorns the promenade is by Hungarian artist J. E. Boehm, who sculpted many of Victorian Britain's worthies. Presumably drawn from contemporary portraiture of the captain, Boehm's Drake stands with squared shoulders, ready sword-hand and left foot forward, scanning the horizon, a globe of the world by his side. The statue was erected in 1884 by the Hoe Committee of the Plymouth Corporation, the city's ancient governing body, at a time when Victorian Britain felt a renewed appreciation for its heroes.

Though Drake can see the sea from this vantage, the statue is set far back from the water, as if to emphasise the way the nation hugged him to its bosom. Victorian Plymouth wanted to commemorate the Drake who circumnavigated the world and defeated the Spanish Armada, not the man who sailed on slaving voyages or thieved his fortune. Most biographies of him have been hagiographies, justifying the less savoury aspects of his career in religious terms as the acts of a Protestant hero against Catholic Spain. To some, he is the founder of England's naval traditions. In Sir Henry Newbolt's 1896 poem 'Drake's Drum', his spirit slumbers somewhere offstage, ready to be summoned by drumbeats to fend off England's enemies (a drum allegedly owned by Drake now hangs in Buckland Abbey, his country seat):

> 'Drake he's in his hammock an' a thousand miles away
> Capten, art tha sleepin' there below?'[15]

Incidentally, the Victorians who erected Drake's statue were at the same time happily tearing down Elizabethan buildings around the old harbour. By their time, the former seacaptains' and merchants' houses had become rack-rented to the poor as wealthier citizens moved to Plymouth's fresher suburbs to the west. Despite the outcry from antiquarians, the once-fine Elizabethan and Jacobean houses were condemned as slums, demolished and their sites redeveloped. Among them was Drake's old townhouse, felled around the time that his statue was erected on the Hoe, where respectable citizens gathered for their promenades.

During the Second World War, Plymouth, a prime target as one of England's most august naval bases, was eviscerated by German bombing. Nearly all the bombs fell on the Victorian tracts of the city directly above the Hoe and west of the old harbour, the Elizabethan streets and houses of which were miraculously spared. It's as though the city's Elizabethan enchantment spared its old heart at the expense of the buildings of its Victorian destroyers. Perhaps this survival of so much Elizabethan architecture in Plymouth's townscape is why it breathes that era especially.

When I visited Plymouth with Dad, we stood and looked Drake up and down, noting the lichen and guano dotting his features. And I couldn't help but wonder if this statue might be where the spell resides – and that if it were toppled, it would be broken.

Would removing Drake from public view, as some have suggested, lead to much of a sea change in the way he is perceived? I thought back to the emphasis that bookseller had placed on his name, the way he had used it rather than the Queen's to denote the bookshop's age. More than any other of the Elizabethan seafarers, Drake has attained trophy-status in Britain's seafaring history; more than any other, the Elizabethan era continues to beguile us with its resetting of England's place in the world as a naval power. Its folklore is sunk deep in the nation's fabric, such that no amount of post-imperial soul-searching will ever expunge it completely.

After we left Drake's pedestal, we retraced our steps downwind of the Hoe and back into the harbour, where a fisherman blundered from a pub doorway as it began to pour with rain. Before we scooted inside, I cast my eye over the quaysides, the cobbles rippling in every direction.

Hardly any are Elizabethan. Most are from later eras, while some are modern but carefully historicised. Regardless, they are the unnoticed detail that completes Plymouth's historic character, a groundscape upon which the Elizabethans might easily walk. In 2019, a proposal to uproot them and repave the Mayflower Steps with sleek new granite setts provoked uproar from Plymouth residents. Council leader Tudor Evans – appositely named – swiftly paused the plans in recognition of the sentiment possessed by the cobbles. And as my father and I sipped the foam from our pints, we talked of how it might be in these stepping-stones, not the statue, that the spell really resides.

'El Draque' ('The Dragon'), was how the Spanish knew Francis Drake. On account of his numerous, lightning-quick raids on the Spanish mainland, they feared him like a mythical beast who burned everywhere he went and seemed to be in all places at once. Today, in Britain, he retains a certain omnipresence; it can be hard to escape the man. As if to emphasise how well woven he is into the national story, trophies of him are to be found in Plymouth, Tavistock, Bankside, Middle Temple – and Oxford.

The light of the afternoon is leaden, and cyclists clatter hurriedly home before dusk. The Divinity School, a long, stone-flagged room built in the fifteenth century as debating and examining chamber, seems to capture Oxford's essence. Over-head, there is a beautiful ceiling of thin lines of stonework, el-egantly bossed and drooping. Fixed under Gothic windows are long, dark woodwork pews, facing one another across the room. Aside from these, the room is bare of furniture except for one other object, displayed here for reasons unknown.

It is a chair in a corner. Stylistically, it sits well apart from the surrounding Gothic. Also of dark wood, but riddled with a classical motif of round arches on spindly columns, the chair looks forbiddingly gaunt at first. Yet a closer look reveals orna-mental scrollwork chased into the wooden beams on which the columns rest. The carving of the segmental headpiece is most elaborate. Flanked by wings, a wind-cherub's purse-lipped face hovers over a metal scroll on which lettering has faded. On either side of the backrest glower faces in headdresses. Below, the openwork body of the chair comprises a series of widely spaced pairs of spindly columns supporting decorative arches. I have seen these arches before, but I struggle to recall where. And despite these decorative flourishes, there is something un-couth about the chair, as though it sits here uneasily. Like a fish out of water, in fact.

In 1662 or thereabouts this chair was fashioned out of

the timbers of the *Golden Hind* and presented to the Bodleian Library by John Davies, Keeper of the Stores at Deptford Dockyard. According to the diarist and naval administrator Samuel Pepys, Davies was a 'very pretty man' who kept a study filled with literature and very good songbooks. His dockyard colleague, Thomas Cowley, had a brother, Abraham, a poet who had been a Royalist during the Civil Wars, served the exiled Charles II in his French court and afterwards studied botany and medicine at Oxford. Between them, they seem to have formed the plan to bequeath this chair to the Bodleian Library, adorned with a verse by Abraham Cowley celebrating: 'This Pythagorean ship (for it may claim / Without presumption so deserv'd a name / By knowledge once, and transformation now)'.

There were plenty of skilled craftsmen in the dockyard who could have wrought the chair, though the joiner's identity is lost to us. An intriguing candidate is Thomas Simpson (or Sympson), the Master Joiner at the Deptford and Woolwich yards: Pepys writes of employing him to make bookcases and chimneypieces, so he was certainly adept at unusual commissions. But why, of all things, should Davies have ordered a chair? Why not something more nautical? Yet there is a delightful tension between the thousands of miles in the wood and the rootedness of the furniture it has now become. And as a symbol the chair, in its way, is every bit as rich as the stories of the *Hind*. For a chair is a symbol of authority, from the highest to the lowest levels of society; Davies perhaps wished to shape a token of her captain's presence.

I then realise it has been odd to spend so much time looking at a chair without sitting down. I briefly consider it, but I hesitate, thinking of the ship's rotten state in the dockyard. Heaven knows how fragile the timbers might be now. The chair's skeletal design seems inherently brittle. And its mysterious, uncouth character seems to emphasise the brittleness of our knowledge of these early transoceanic voyages. It's ironic, also, that the only sizeable, proven piece of an

Elizabethan galleon reposes in landlocked Oxford. You might say that it's now as distant from the sea as we are in time from its sailors.

I let my gaze rest upon the chair a moment longer, then quit the Bodleian for the lamplit streets, where the cold of the season catches the tonsils. Silhouetted by starlight, the delicate sandstone pinnacles are crisp as though new-cut. Such fine detail could not survive on the coasts, exposed to the salt and damp that the sea brings; and the still chill in the air here would never be felt on the quayside, at the edge of a vast, living tumult of water. Gripped between these contrasts I attempt, without much luck, to unpick my impressions of the chair. How close had I really come to the *Hind?*

Then I recall where I've seen that persistent arch motif. Arches of that sort had decorated the replica of the ship, lining her fore- and half-decks; they are echoed in the drawings in the only Elizabethan ship-blueprints that survive, those of Matthew Baker which were collected and preserved by Samuel Pepys though his dockyard contacts.[16] A tantalising theory swims into view: that this chair, incorporating the ship's main decorative detail, carved with tokens of its voyage (headdresses, blowing wind), is in fact a kind of occult portrait of *Golden Hind.* Perhaps it is. But all we can certainly say is that it reveals just how far the Elizabethans have become part of the national furniture; and that what remains of this oceangoing vessel now nestles far inland, a carrier of trophies transformed into one.

6. The Drake Chair, 1875 photograph.

Rope

The Ropery

TRAFALGAR SQUARE AND EMBANKMENT GARDENS,
LONDON; CHATHAM DOCKYARD, KENT
STUART AND GEORGIAN

The length of junk was dragged from the cart and heaved onto the stone-flagged courtyard. It lay there in huge, blackened coils, reeking of heavy use and heavier sea. It had seen much, this old length of ship's rope, called junk when it ceases to be useful. It had gone hoary with ice all the way through the roaring forties, way south of the Cape, while under tropic suns it had spiralled with molten tar. For most of its life it had twanged and tautened over the Atlantic. Now it lay in the workhouse courtyard, awaiting its resurrection.

First, the brawnier inmates broke down the rope with picks and shovels, reducing it, near enough, into a series of fist-sized cross-sections of itself. As they did so, more odours were freed from the junk, called 'balsamick' by one approving writer of the time.[1] These pieces were brought inside to where the older, more infirm inmates were gathered at benches in the workhouse interior. With their bare fingers they sought to loosen the hard-packed fibres of each piece of rope, perhaps slipping a nail under the cut edges to start the unravelling. The sheer quality of the ropemaker's work made this an abominable task. Gradually, though, the pieces began slowly to sunder into tarry fragments of hemp, chafing painfully upon the hands.

Early one spring morning I stand at the top of Whitehall. Behind, Trafalgar Square is blessedly empty and quiet. I chose this crowdless hour very deliberately to admire the Admiral atop his column. He's some fifty metres above me, so I put the binoculars to my eyes and sweep the sky that frames his silhouette, focusing in as best I can. Suddenly I find his face in my eyepieces, one eye unseeing below the bicorne hat, the other unblinking, an alarming close-up no-one save his intimates ever had.

Horatio Nelson was born in Norfolk in 1758, in a place where, compared to the huge vistas of sky and sea, land seems incidental and uncertain. Under the aegis of his cousin he was admitted to the navy and gained his first command when he was twenty years old. In the West Indies he excelled at frigate actions. Later, he cruised the Mediterranean, resisting promotions into larger vessels because these would place him at a further remove from action. In 1798 he ran down a French fleet anchored in Aboukir Bay and, threading his squadron amongst shallows and sandbanks, utterly destroyed it. He disobeyed his superiors at Copenhagen in 1801 and forced a truce out of a superior Dutch foe. Finally, in 1805, he led in *Victory* the British fleet that annihilated a Franco-Spanish fleet at Trafalgar, saving England from invasion at the cost of his own life. A French seaman shot him on his own quarterdeck.

In Nelson there is a great tying-up of the loose ends of a chaotic eighteenth century, one which saw periods of naval reversals and advances against a shifting cast of foes; with him, the Royal Navy finally cowed its main rivals and reigned supreme over the sea. Ever since, he has been England's naval figurehead. And this column could be called England's premier naval monument, raised a few decades after the battle which claimed him. Fittingly, his statue was sculpted by the son of a ship's figurehead carver, Edward Hodges Baily. In my eyepieces I savour the familiar details: sleeve pinned to weskit; hat athwartships; hand upon sabre-hilt; pose both noble and poised. I find the admiral everything a naval hero should be.

But the column is not just a monument to Nelson; with its relief panels depicting the key battles made from melted cannon, it is also a monument to the navy he represented and helped shape: a naval ideal of warm wood, fellowship, talent, courage and devastating cannonry. And a monument to sail. In my eyepieces, I see that he's not alone on that great pedestal. Beside him is a stone coil of rope.

Fittingly so, for the age of sail was really the age of rope.

However finely drawn and carpentered a ship's lines, she was static until her ropes were hauled to make her a vehicle. Ropes brought the ship to life, furling or unfurling the sails, positioning them to tempt the wind, and trimming them to catch the best of the gusts. Ropes helped to steer the ship, joining wheel to rudder. Ropes held things in place: the shrouds and stays, for instance, tied the masts securely to the decks, while ropes entwined into anchor cables held the ship to the seabed. As well as these primary functions, ropes had all sorts of ancillary uses aboard: swaying up timber beams for repairs to masts or spars or yards, craning aboard and manoeuvring the heavy iron guns.

A length of a rope seems a humble thing, the antithesis of the majesty of a warship, yet ropes were central to naval operations. No wonder, then, that sailors are synonymous with knots, or that ropes adorn virtually all monuments to or depictions of seamen. For centuries, rope more than sails or ships was a kind of artistic shorthand for the sailor. But rope is the least enduring of all a ship's parts, far more perishable than the ship's timbers or its ironmongery. Like the seafarers whose hands manipulated them, ropes wore out quickly. Now that the age of sail has long vanished, both seem almost irrecoverable. As the historian N. A. M Rodger puts it: 'few lives seem to have vanished so completely, in so short a time, as that of the square-rig sailor'.[2]

Purposely or not, Nelson's Column commemorates this. For what is the column but a symbolic mast, thickened and enriched with carving? In height, it's comparable to the masts of HMS *Victory* (lower than the main, higher than the fore and mizzen). And in keeping Nelson aloft, the column commemorates the lost activities there that were the quintessence of sail – the loosing, reefing and furling of the sails, the scramble to the topgallants (the topmost parts of the masts) to do this – unsafe at the best of times, positively perilous in a gale. But, in the end, Nelson's Column is a monument to one man. We must look elsewhere for a truer monument to the square-rig sailors who have vanished so completely. Fortunately, the stone coil of rope beside Nelson hints at where this monument can be found.

Lines figure heavily in naval affairs. We speak of a ship's 'lines', her silhouette and profile in plan and cross-section. We speak of ships of the line – large warships whose task it was to form lines of battle; we speak too of the lines of supply that supported the Royal Navy in its operations across the world. Ropes are the embodiment of these lines in every sense and, today, serve as an instrument to recall the golden age of the Georgian navy. Even the behaviour of a rope seems to reflect that of a navy: the continual veering between slackness and tension, the discipline required to keep it coiled and properly strung, the hopeless disarray into which it could fall at a moment's notice.

'The Straits [Mediterranean] cables are the best . . . the next, the Flemish and Russian, the last, ours . . .'[3] Writing in the 1620s, Sir Henry Mainwaring's assessment of the quality of British ropes reflects the mixed fortunes of British sea-power under the early Stuarts. The Tudors had left the makings of a modern navy: fledgling administration, a fleet of nimble new ships, new gunnery techniques and the legacies of the mercurial seamen who had thrust English horizons westwards. But it was an ensemble kept tensed and effective by Spanish enmities. When peace came in 1603, the navy lapsed into a slackness that was the antithesis of that vigorous fleet inherited from the Tudors.

I leave Trafalgar Square and make for the Strand, dropping down to the river to see another monument, one which speaks unintentionally though powerfully of a lesser navy. If seeking to understand the history of the Royal Navy, it is perhaps a truer place to start than Nelson's column.

In 1626, the York water-gate was built as the river entrance to York House, Thames-side residence of the Lord High Admiral of England: George Villiers, Duke of Buckingham. As of 2021, its context is much changed. Now shorn of the river walls it once pierced, far from the water to which it was once a portal, the water-gate is sandwiched between a park and a narrow avenue, Watergate Walk, which runs adjacent to it after springing eastwards from Villiers Street. It happens that I once spent much time – and much of my student loan – in the lee of this water-gate, sitting at one of the tables of Gordons' Wine Bar which have annexed the avenue. Back then, it seemed to me like just another chunk of masonry that had fallen into the present, rather than a rare and curious sea-folly.

I revisit on a day when the bar's shut; the tables are mothballed and no drinkers knot the avenue. Inspecting the landward, once-private side of the water-gate, I think fleetingly of the sour house red, the copious cheeseboards, the too-powerful French cigarettes like Gitanes and Gauloises we smoked here solely for their cachet. Now I study the water-gate with older eyes. Three arched openings are divided by pilasters, keystoned with cartouches and surmounted by Buckingham's motto: FIDEI COTICULA CRUX ('the cross is the touchstone of faith'). I look through the central archway but the river can no longer be seen.

Upon accession, King James VI and I promptly made peace with Spain. He was a writer instead of a fighter. His reign saw campaigns against pirates rather than rival navies. Money was spent on the dockyards, but comparatively little on new ships, which were usually built of weak greenwood, and very little on seamen. Corruption was endemic: in these slack years of peace, the navy was seen by its administrators as an opportunity for embezzlement, not excellence. In 1604, the master shipwright at Chatham, Phineas Pett, built the *Resistance* entirely from 'defective' or 'missing' naval materials in a private yard downstream.*

---

* Incidentally, *Resistance* was of similar tonnage to *Golden Hind* and would be captained by Henry Mainwaring in piracy against the Spanish until 1618, when he was pardoned by James. Mainwaring was in some ways a successor to Drake and his privateering ilk. Although he died poverty-stricken in 1653 after a chequered career of service to the Crown, his *Seaman's Dictionary* is a valuable porthole into the workings of an early seventeenth-century ship.

To be fair to James, he occasionally looked up from his books to examine the dockyards. Inquiries of 1608 and 1618 laid bare the scale of the problem, and a little effective action was taken. In the latter year he appointed his favourite, George Villiers, as Lord High Admiral. Buckingham was keen to better the navy and energetically pursued a campaign of reform. He was equally keen to use it to strike sparks against Spain after the breakdown of marriage negotiations between Prince Charles and the Spanish Infanta. Elizabeth's reign loomed large over Buckingham's plans, which were eventually manifest in the form of an expedition against Cadiz in 1625. It seemed to imitate various campaigns by Drake against that important Spanish harbour, but Buckingham's attempt was a disaster: rotten ships and rotten men achieved little, and returned home in such a pitiful state that they resembled the survivors of a botched circumnavigation. The next year, licking his wounds at York House, Buckingham built his water-gate.

The Jacobean navy could not be relied upon to protect English shipping, or even patrol the Channel. Rival navies sailed unhindered through its waters. On the same visit in which he saw the remains of the *Golden Hind*, the Venetian ambassador also noted: 'we saw what may be called a new arsenal, begun by the India Company and on the same day passed alongside of two immense ships called "*the Sun*" and "*the Moon*", which are completely found for the India voyage, with all their hands and munitions. They really looked like two well appointed castles'.[4] In a departure from those Elizabethan joint ventures, private shipping fended for itself, unwilling to become entangled in bungled naval ventures. Buckingham was forced to rely on a compulsory levy of private ships for his next venture, an expedition against France, but was furnished with only meagre vessels, poorly crewed; the result was a massacre of the English.

In 1628, Buckingham was assassinated. A few years later, relations with France and Spain were smoothed over. Subsequently, King Charles I took a close interest in the navy. In 1631, he toured the dockyards and ships, going 'aboard every ship and into almost every room . . . and into the holds of most of them'. Under Charles I, the Ship Money tax revived the dockyards and funded the construction of glittering goliaths, of which the *Sovereign of the Seas* (1637) was the masterpiece. She was the first English warship to mount 100 guns (102, in fact). Although ill-suited for patrolling the Channel and protecting British shipping – the most pressing naval need – Charles's ornamental warships were an effective declaration of intent, and Britain began to recover some naval prestige, albeit without addressing the problems of corruption and mismanagement that plagued it at the root.

I circle around into Embankment Gardens, a park made on land reclaimed from the river which once swished against the water-gate steps. I want to see its riverine face – and what a face. Advertising a ducal house to the Thames, it's much grander than the side I sat against as an ignorant student. From this side, it takes the form of a central archway flanked by smaller arched windows framed within thick columns and mighty keystones above. Virtually everywhere, the stonework is rusticated – deliberately coarsened rather than smooth. Above, Buckingham's arms are held in a broken pediment flanked by decaying lions, while downturned escallops festoon the frieze below. These last were considered sufficient to depict his admiralship. There are no traces whatsoever of ropes.

Grotesque in aspect, long detached from the river which gave it purpose, the water-gate looks exactly like an architectural folly built to evoke a lost past. And this it certainly does, but in such a way as to embody the detachment and gross corruption of the early Stuart navy. It's a relic which forces us to confront a time when the Royal Navy was barely surface and no depth, sadly lacking in the simplest degree of seamanship, a mere façade compared to the Elizabethan substance which had preceded. Its archways may no longer frame water, but it's a useful porthole onto a naval nadir.

I drive south-east out of London through pylon-trampled Kent, on the road to another naval monument, if a whole anchorage and its fringes may be so called. I seek a lost fragrance of sea-faring Britain, the reek of hemp, sisal, tar now absent from our quaysides. In the windscreen, freeze-frames of unspoiled countryside skip and segue with railways, roads and wirelines bearing down on the Channel ports. I peel off the motorway at the exits for the Medway, a river which flows conspicuously through naval history from Drake's time onwards. It has much to tell about subsequent evolutions of the Royal Navy, the scene both of its greatest defeat and one of its greatest triumphs.

Following Charles's beheading in 1649, the Dutch displaced the Spanish as the main seaborne foe. While Spain was from this point a waning antagonist, the rising maritime power of the Dutch was developing a virtual monopoly on trade with the Far East and challenging English interests in the Americas.

In the first Anglo-Dutch war of the 1650s, Cromwell's Protectorate fought with mixed results, though in doing so built up a pugnacious and well-administered navy. That war ended in stalemate; low-level hostility between the two nations simmered over the Restoration, when in 1660 Charles II returned to rule England in ships reclaimed (and renamed) from the Puritans. Charles and his brother James were unprecedently well informed on naval affairs and James was a skilful and gallant Lord Admiral. Under the merry Restoration court, naval corruption persisted but was valiantly fought by the likes of Samuel Pepys, who documented all in his famous diary. Amongst the skulduggery recorded therein, one entry caught my eye: in August 1663 Pepys recorded how, in trials at Chatham, English ropes now compared favourably with those made from Russian hemp.[5]

After parking up in a leafy glade, I walk down a lane under the airborne tang of burning wood. Outside a pub, I come across an old dinghy sawn into thirds, the bows having been upended as a makeshift shelter, and the midships and stern halved to make benches. They also halve the dinghy's stern-painted name: *Arethusa*. Further down, I spy something even more arresting: one-third of an enormous wooden rudder, surely once that of a man of war, re-hung as someone's massive gate-hinge. I'd like to meet this whimsical person, for they must also be responsible for the curious octagonal lookout tower rising above. Opposite, the gates to the castle are closed, but no matter. I reach the riverbank having fallen hard for this Kentish village of mellow brick and weatherboarding.

Upnor lies on the north bank of the river Medway. On a map, the river's estuary is no smoothly splaying trumpet-bell. Instead, it looks more like a set of bagpipes: a comparatively narrow passage between Sheerness and the Isle of Grain broadening into a lung of channels, islands, islets and marshes, a riddle to foreigners who would seek to navigate it without pilotage. Little tributaries stem from this body of water like reeds and pipes: eventually the estuary narrows to a funnel between St Mary's Island and Hoo. Here, the Medway regularises. Follow it around St Mary's Island and you come upon a long, linear reach which ends at Chatham, where the river pivots away inland.

My trainers are inadequate for the terrain, and squelch deeply into the waterlogged track along the shoreline. I walk for a while upstream, to get a better view of Chatham Dockyard and its sublime Georgian frontages which line the river and speak of a lost maritime strength. Soon my way is blocked by thin strands of barbed wire denoting a military base, so I turn back towards Upnor. Only small yachts are moored in the river's central road. Over the mudflats unveiled by the ebb tide, a lone hovercraft buzzes aimlessly. Behind me, smoke issues from a huddle of houses, while from this muddy shore I do not see the sea, only the abrupt bend of the river.

From the Tudor period, this Medway reach was a winter harbour for the English navy. Distant from the sea, sequestered from hostile fleets and winter gales, it's a perfect place for keeping warships. But it became more than just a wintry anchorage. Under Queen Elizabeth a dockyard was built at Chatham for the construction of warships under her new navy and the maintenance of the existing fleet. The first ships recorded as being Chatham-built were the *Merlin* and the *Sunne,* names with which to further conjure the strange magic of Elizabeth's reign. A castle was also built upstream from the dockyard at Upnor, able to cover the approaches with gunfire in the improbable event that an enemy should successfully pierce the complex estuary.

At least, it seemed improbable until June 1667, when a Dutch squadron mounted a raid on the Medway, navigating the difficult approaches by using traitorous English pilots, then coming upon four of the five English flagships moored between Upnor and Chatham. Only skeleton crews occupied the *Royal Oak, Royal Charles, Royal James* and *Loyal London* and the surrounding fleet. England and the Netherlands were in the throes of their second war; the previous year, a series of ferocious naval battles and the Great Fire of London had exhausted the Treasury and forced the fleet to be mothballed in the Medway.

That summer's day, the Dutch burned *Royal Oak, Royal James* and *Loyal London* – costly bonfires of timber, tar, canvas, cordage, paint, gilt and saltpetre. They sent a prize-crew aboard *Royal Charles*, led by a renegade Irishman, who unmoored her and absconded. By then the tide was on the ebb, and to get the massive warship over the shallow channels the Dutch seamen audaciously heeled her over to one side. Of this, Samuel Pepys wrote: 'they did carry her down at a time, both for tides and wind, when the best pilot in Chatham would not have undertaken it'.[6] Then they roped her to a Dutch warship, towing her across the North Sea and into the little Dutch port of Hellevoetsluis, where she was drydocked as a trophy in a strange reflection of what had happened to *Golden Hind* nearly a century beforehand.

Of all the enemy incursions into England, this was perhaps the most humiliating. True, the medieval coastlines of the south and east had greatly suffered in the Hundred Years' War; for instance, two French burnings of Melcombe in Dorset irrecoverably destroyed its prosperity (and, unhappily, the first case of the Black Death was also recorded there). True, in 1595 the Spanish landed unopposed in Mount's Bay, Cornwall, and burned the fishing-ports. And, true, under the early Stuarts, Cornish and Devonian ports were easy prey to Algerine pirates, who carried off many of their inhabitants as slaves; my home port of East Looe lost eighty men in 1626, and sixty-nine a decade later.[7] But these raids, though grievous, were local affronts.

By contrast, the Dutch raid on the Medway was a national humbling, made all the worse because the navy was beginning to recover its strength. It had performed creditably in the second Anglo-Dutch war of the 1660s, which was triggered by the English capture of New Amsterdam in the Americas (or New York as it is now known). Subsequent North Sea battles were infernos of salvoes between vast fleets: the Battle of Lowestoft of 1665 could apparently be heard in Hyde Park.[8] This was a navy of gunsmoke, thunder and gore, of long lines of ships pounding one another with shot before breaking down into a melee. This was a navy in the ascendant after its Jacobean and Caroline nadir. As such, the Dutch raid on the Medway of 1667 was all the graver a humiliation.

Today, Upnor is untroubled. The logsmoke in my nostrils is like the faintest echo of those burning flagships. And, anyway, flagships are only the figureheads of the whole; powerfully symbolic as they may have been, they could be rebuilt in just a few years. Had the Dutch managed to reach Chatham, just upstream, a far graver blow would have fallen. Had the dockyard there been destroyed, British naval strength would have been set back for a generation. And it's to Chatham Dockyard that I now go to peer under the bonnet, as it were, of the navy in its prime.

One last circuit around the huddle of houses and along the Medway shore. I get the feeling of trauma mellowing to curiosity with the long swishing of the centuries; this tiny Kentish village gives an unusually acute sense of great reversals and their subsequent insignificance in the grand scheme of things. I get the feeling that if I raised the Dutch raid with passing residents, they'd only respond in smiling silence. On the ebb, the river shivers unhurried down to the estuary; slowly breathing woodsmoke, the village hangs as though in a kind of golden suspension. Here, past events feel slackly coiled, rather than held in tension still.

On my left is a wall which seems to run for miles. It's built of mottled, flesh-toned brick, sparsely detailed with unadorned pilasters and crowned with a triangular brickwork ridge. The wall is taller than me, for it was designed to keep out prying eyes, foreign spies and other ne'er-do-wells. Its plain design has the effect of discouraging speculation about what lies on the other side. Yet it hides a triumph.

I walk the wall with a little scepticism that I will find some of the greatest monuments to the age of sail on the other side. For, as a town, Chatham seems to be recovering from rather than basking in its seafaring past. The Royal Dockyard closed in 1984. It had been the engine that had swelled a rural village to a sprawling seaport, which in 1798 could be described by the antiquarian Edward Hasted as 'long, narrow, disagree-able, ill-built . . . the houses in general occupied by those trades adapted to the commerce of the shipping and seafaring per-sons'.[9] Now only the first half of this description holds water.

The wall runs along the dockyard's northern boundary. Soon I come upon the main brick gate of 1722, gaunt and folly-like with the glistening arms of George III squared above the portal. The yard boatswain once lived in one of the towers, I think. But for some reason I'm not allowed through it, and must walk for longer in the lee of the high dockyard wall, along a narrow road smoggy with traffic, until I come to the main visitor entrance at the north end. Here unfolds the sort of vast space characteristic of redeveloped military-industrial sites. After its closure, much of the north portion of the dockyard – dating to Victorian and modern times – was infilled, razed and redeveloped, bound in the chains of roundabouts and nonde-script buildings. Across the Medway, Upnor still smokes attrac-tively in the distance.

If the Elizabethans represent a coming-of-age of English seafaring, then the Georgian age must be its prime. It is in the

coils of this age more than any other that we are still entangled as a nation. Perhaps one of the reasons for our affection is
that this was an age which loved its navy. Victories were deliri-
ously celebrated; defeats were felt bleakly, keenly. At the root
of this lay the belief that the navy was a kind of 'wooden wall'
which protected England from invasion (in the days when England required such walls against its foes). In some places it was
much-patched, loosened and faintly archaic in design, while for
other stretches it ran battered but defiantly intact, while for yet
others it had been rebuilt to splendid new designs using the best
oak the country could muster. As the Medway raid shows, this
wall was never perfect, but it steadily earned its nation's affection by successfully defending it from seaborne attack.

At the Glorious Revolution of 1688, the Dutch were superseded by the French and the Spanish as the main seaborne
foes. The next half-century or so saw a continual tensing and
slackening of hostilities with France, entwined with invasion
scares and inconclusive battles fought in faraway waters, until
the totemic year of 1759. This came to be known as the *Annus
Mirabilis* because of a series of British land and sea victories
culminating in the Battle of Quiberon Bay that November. Dar-
ingly, a British fleet led by Sir Edward Hawke chased a French
fleet into those notoriously reef-strewn waters in stormy
weather and sank not just the French warships but also the invasion of England which they were to have facilitated. Perhaps
another reason for our addiction to this era is that this was so
obviously a talented navy – we all look back with affection on
our prime.

By this time, the fleet's mainstays were large warships
mounting copious gunnery along their flanks. In contrast to
those nimble Elizabethan galleons, which would chase down
an enemy ship, let fly with their cannon, then scurry away to
reload, these Georgian warships fought in lines of battle: they
would form a line facing the enemy and loose broadside after
broadside in a bout of crushingly heavy fighting until one or the
other was destroyed. Unlike the slightly smaller class of frigates, the job of which was to cruise on a range of offensive and

defensive duties, ships of the line really had only one task: to issue as much firepower as possible and to take as much in kind without sinking. This tactic had been formalised in the Cromwellian navy, but had been much evolved by Georgian captains such as John Jervis, Earl of St Vincent and Nelson, who developed tactics to break enemy lines. These ships of the line came to symbolise the pinnacle of the Georgian navy. To John Ruskin, looking back on this time from the 1850s, a ship of the line was 'the most honourable thing that man, as a gregarious animal, has ever produced . . . into that he has put as much of his human patience, common sense, forethought, experimental philosophy, self-control, habits of order and obedience, thoroughly wrought handiwork, defiance of brute elements, careless courage, careful patriotism, and calm expectation of the judgement of God, as well can be put into a space 300 feet long by 80 broad'.[10]

Only one ship of the line survives today:* HMS *Victory*, drydocked at Portsmouth. Although little of her original fabric remains, her outlines are those upon which Nelson paced during Trafalgar; her quarterdeck still the one on which he was shot in the heat of that battle. She was built at Chatham to designs by the influential shipwright Thomas Slade and launched in 1766. And her lines owed much to the French warships which Slade was instructed to copy, they being acknowledged as superior to English vessels in design but not in the quality of their construction or their seamanship.

When I saw her at Portsmouth, I found her smaller than I had expected, with tubby yet graceful outlines like those of a mallard on a pond. Immediately I could see – or perhaps I had been conditioned to see – that she was more than just a warship. Indeed, her stern prickles with architectural features, forced aslant by nautical necessity: lopsided sash windows, for instance, with their glazing bars skewed like hashtags. Her sides are lined with three rows of gun-ports, but curiously these

---

* There is, however, another survivor from the later Georgian navy: HMS *Trincomalee*, a frigate built in Bombay in 1816 and now moored in Hartlepool.

did not seem the standout feature; she appears, to borrow the words of Frederick Chamier, a young Georgian midshipman, like 'a kind of elegant house with guns in the windows'.[11]

At least, that is my impression of *Victory* now, but Chamier was actually speaking of what he expected to find when he arrived at his first posting, the thirty-six-gun frigate *Salsette*. The reality was quite different: 'Ye gods, what a difference . . . here were the tars of England rolling about casks, without jackets, shoes or stockings . . . the deck was dirty, slippery and wet; the smells abominable; the whole sight disgusting . . . I forgot all the glory of Nelson, all the pride of the navy . . .'[12]

7. *Victory*, stern detail. Naval architecture in several senses

Such ships were built and refitted in the Royal Navy dockyards. And they themselves were dominated by lines and walls – vast extents of timber, brick or stone, forming storehouses and offices, roperies and docks, enclosing vast tracts of the ground required as the dockyards expanded to accommodate warships growing in size and multiplying in number. They were great, pre-Industrial behemoths, easily the largest consumers of raw shipbuilding materials and the largest employers anywhere in the country. Often whole towns, like Chatham, utterly depended on them. In their size and sheer presence within their still-agrarian settings they anticipated the Industrial Revolution. And they were the terrestrial equivalents of the fleets that lay potently out of sight.

Portsmouth lays claim to being the oldest Royal Dockyard, having been used since late medieval times to repair and harbour the King's ships, lying conveniently on England's central axis. To the west, Plymouth is one of the newest, having been established after the Glorious Revolution of 1688 to counter the emerging French menace, France having usurped the Netherlands as the main seaborne threat. It would remain so, in fits and starts, all the way up until Trafalgar. Other dockyards lay at Deptford, Woolwich, Sheerness, Pembroke and Chatham.

During the Dutch wars, Chatham's eastern location made it the natural naval focus. But with the switch in enmity to the French, Chatham became less convenient; moreover silt, that perennial enemy of ancient anchorages, was gradually beginning to afflict the Medway. And ships were growing in tonnage – ships of the line were struggling to navigate upriver to Chatham, except at spring tides when the water was highest. Consequently, from the first half of the eighteenth century, it was decided that Chatham should be a yard for shipbuilding, in the main, rather than the dedicated maintenance of the fleet.

This meant that the character of the yard was defined less by the frenetic crewing, victualling and refitting of a live fleet than by the slower pace of building a new one. And in the eighteenth century it was sheltered, figuratively at least, from the front line of naval operations in the west: the blockading of the French fleet in Brest or Toulon, the cruising of a Western Squadron covering the Western Approaches, the traffic of frigates and warships to the West and East Indies on longer-range missions to defend the British colonies there or annex new ones from rival powers. All these placed heavy demands on Portsmouth and Plymouth, with their waterfronts thick with the frenzy of maintenance, repair and supply. That at Chatham, on the other hand, was characterised by half-built ships in various stages of advancement, slowly building towards completion. And in that totemic year of 1759, the keel of *Victory* was laid down in this yard.

Ships of the line were christened for their formation in battle, but they were well named in another sense. For the ships themselves were dominated by a bewildering family of lines, a sublime complexity of ropework. Bolt-ropes for reinforcing the edges of the sails. Bowline for bracing the sails sideways in adverse winds and stopping them from shivering. Halliards for hoisting or lowering them. Horses and stirrups for sailors to tread while manipulating sails on the yards. Shrouds for tying the masts to the decks. Preventers for supporting other ropes. Tiller-ropes, untarred for smoothness, joining wheel to rudder. Mainstays for securing the mainmasts. Over thirty miles of rope ran through the workings of a typical ship of the line. Not only did these mainstays of the navy project English sea-power over great distances, but they also coiled great distance within them.

At the core of Chatham dockyard is a precious, near-complete Georgian shipbuilding complex. With ships of the line now lost to us *en masse*, visiting Chatham, with its fleet-like ensemble of the different dockyard buildings, is the only way to gain a sense of the magnificence of the Georgian fleet and the plethora of crafts and trades on which it depended. For naval operations have ceased completely here, allowing visitors to roam about the made-over storehouses, the stoppered slipways, the frozen machine shops. Other dockyards are either still operational – such as at Plymouth or Portsmouth – or are very heavily altered so that only traces of their Georgian splendour linger.

For they were splendid – a fact not lost on contemporaries. Edward Hasted, who took such a dim view of Chatham town, saw in the dockyard 'many elegant buildings . . . which well become the opulence of the nation, and the importance of the navy'.[13] Visiting earlier, in the 1720s, the merchant, writer and polemicist Daniel Defoe wrote of the dockyard buildings that they were 'indeed like the ships themselves, surprisingly large, and in their several kinds beautiful'.[14]

In a sense this was quite true, for earlier the Navy Board, ever thrifty, had issued an instruction for new dockyard buildings to 'make use of old timbers from ships being took down'.[15] In Chatham's Sail Loft of 1723, ribs from what was probably a Stuart warship were reused as structural pillars. Elsewhere, ships' knees – curved pieces to support the decks – were reused as supports for roof structures. Only rarely can the ships be identified. In 2012, timbers under the floor of the Wheelwrights' Shop were identified as those of *Namur*, a Chatham-built Trafalgar veteran whose timbers were most likely interred there as a commemoration of her career.[16] But she is an exception. Most of the timbers are from now-nameless ships, half-resurrected as these buildings. In this literal sense, the navy is survived by its dockyards.

Approaching the dockyard from the north, through the vast car parks made over infilled drydocks, I pass a lone Mast Pond of 1697 (for seasoning the great pines in brine), retained as a memory-pool. Nearby, vast sheds of later ages screen the river, under which iron and steel warships reared on their slip-ways. At right-angles to them is an older, zig-zagging roofline of a lower, Georgian range. It was the Mast House and Mould Loft, for shaping the masts on the ground floor and shaping ships on the open drawing-floor of the loft. Here, embryonic ships' designs were drawn out to life-size on the floor to make moulds for the cutting of their timbers. And in this and other of the dockyard buildings, in their proliferating timber and unusual proportions, there is a feeling of great closeness between the ships and the architecture.

Beyond the Mast House is a muster-ground fronted by the gables of brick smithies and steel sheds, between which slivers of river may be glimpsed. Museum warships – of sail, of steam, of nuclear power – are exhibited in old dry docks, shaped into the negative image of the hulls they were to receive. Their stepped stone sides are like the raked seating of a theatre. I suppose all this *is* a kind of theatre – and we do speak of theatres of war – but only for the senses of sight and touch, the sounds and smells and tastes of this place having long drained away with its decommissioning in 1984. Indeed, the dockyard today is largely empty of people, save a few meandering tourists. The young Frederick Chamier would find it conspicuously silent. Once it would have been thick with milling personnel: the shipwrights, aristocrats of the dockyard, loftily inspecting the progress of their creations; the anchorsmiths, forge-reddened and slightly tipsy from the eight pints of strong beer they were allotted; the ropemakers, dextrous spinners, layers, hatchellers, coughing the hemp particles out of their lungs.

Moving on south, past handsome quarters for the dockyard management and a stand of lush trees, I spy in the distance the object of my visit, a series of looming brickwork frontages whose slimness – a mere two or three bays wide – belies their extraordinary length, unseen from this angle. It is like looking at the small, circular end of a mast until you shift to obliquely see the whole.

I step to one side and see a wall vanishing south, four storeys high and punched with uniform, rectangular window openings like an endless phalanx of gunports. I go alongside the north end and feel like a dinghy sidling up to a flagship. I expect to find the stillness of the other dockyard buildings, but there is a thread of vitality in this one. From within come the muffled sounds of machinery and slithering cordage. Aromas of hemp and sisal hang about an open doorway. I place a palm on the patinated brown brickwork. My nostrils flare.

To make rope, you must first lovingly comb tangled hemp upon the hatchel. This means carrying the bales from the Hemp House – in which cool darkness they have been stored since they arrived from Russia – to the Hatchelling House, a lovely word which I keep misreading as hatchling house. Although the words are etymologically unrelated, there is truth in the error. For the hatchel was a wooden board bristling with iron spikes, through which the hatchellers dragged the knotted hemp fibres to straighten them. It was the first stage in hatching a new rope.

From there you must take the straightened hemp fibres to the Spinning House, where they can be twisted into yarns. You hook the fibres to a spinning frame, then turn it as the spinner slowly walks backwards, watching as the yarn forms under the guidance of his left hand. Simultaneously he will add new hemp using his right hand, allowing the yarn to reach the correct length.[17] Once the yarns are formed, you bring them to the White Yarn House, where they are slowly drawn through a large brass tar-kettle suspended over a fire. When they are tarred, you take them to the Black Yarn House to cool and dry out.

Finally, you bring the black yarns to the ropewalk. You tie them to bobbins at one end and at the other to a rotating hook mounted on a wheel frame. You push the frame along the laying floor, drawing the black yarns from the bobbins and twisting them together into strands. Then you stretch the strands along the laying floor, mounting them upon trestles to keep them at constant height, before winding them together to form rope. Heavy, smooth, spiral-ribbed rope.

Both Hemp and Hatchelling Houses are nondescript buildings, with nothing about their architecture to suggest their purpose. The Spinning House and the ropewalk, however, make their function plain. They are impossibly long because ropemaking requires length. A new rope's length was determined by the ground available for it to be made or 'laid', which had to run flat and even for the distance. And early ropewalks were just that – open tracts of walkable ground for ropelaying. Being long and linear, they frequently became streets when their original use ceased, and the proliferation of English streets named from ropes or cables – Cable Street in east London, the Ropewalks district in Liverpool – attests to the frequency of ropemaking from the Middle Ages onwards.

But in rainy Britain, uncovered ropewalks risked disruption to their production. So, a distinctive typology of ropehouse began to emerge, usually simple vernacular buildings that were dramatically long, like cottages or barns surreally elongated. Although rope acquired connotations of the sea because of its centrality to animating a ship, its uses on land spanned everything from agriculture to execution – ropemaking was not confined to the seaside. In rural settings, the length of these set-ups was often comparatively modest; for instance, the surviving village ropewalk at Aston Sandford, Buckinghamshire, is scarcely longer than a church nave.[18] And in pre-industrial Britain, there were few areas of terrestrial life – except, perhaps, architecture – where ropes took particularly heavy loads.

Naval ropehouses were the longest buildings in the country. They had to be. Not only were they required to lay cables of 120 fathoms, but also the naval appetite for rope was prodigious. As mentioned above, ships of the line each required over thirty miles of rope to function, and in the eighteenth century there were hundreds of them. Originally, the ropehouses at Chatham were 1,120 and 1,160 feet long.[19] Of this, 720 feet was required for laying the 120 fathoms, with space either end for machinery, storehouses, offices and other ancillary needs. And naval rope had to be much thicker than any required on land, for it must contend with the vast forces of the sea. Greatest amongst them were the anchor cables, a mighty 24 inches in diameter, formed by 220 sweating ropemakers.

A ship depended not just on the quantity but the quality of its rope. To William Falconer, compiling his *Universal Dictionary of the Marine* of 1769, 'we may venture to assert, without violation of truth, that many good ships have been lost only on account of a deficiency in this important article'. Furthermore, ropemaking quality was directly influenced by the skill of the spinner or layer; it could not be done by anybody. Falconer goes on to say: 'a cable ought neither to be twisted too much or too little; as in the former state it will be extremely stiff, and difficult to manage; and in the latter, it will be considerably diminished in strength'.[20]

This work was psychologically and physically taxing. The exertions required to twist together the mammoth anchor cables, the constant abrasion of fibre on hands, the inhalation of hemp dust that could lead to respiratory problems. Rope-makers knew their worth and frequently went on strike when changes to their quite generous terms of employment were proposed by the Navy Board. In 1729, those at Chatham staged a walkout; in 1745, they walked out again over the proposed increase in the quantity of servants attending them, which, convolutedly, meant an increase in their work without a corresponding increase in their pay. Interestingly, the 1745 episode is consistently described as a 'mutiny' by the authorities; the ringleaders were imprisoned aboard the *Royal Sovereign* laid up in the Medway.[21] But rope was too crucial to the yard and terms were soon reached.

The ropemakers might have found readier cause for mutiny in the state of their workplace. The Chatham rop-ehouses were old and much-patched even when Defoe saw them. They had originally been built in 1624, the year before Buckingham's disastrous expedition to Cadiz, and took the form of two extremely long, single-storey timber sheds open on one elevation, one for spinning, the other for laying. As a result, they were not particularly secure: in 1719 they were thought to 'lye naked and [liable] to the ill designs of every desperate sly villain and the bolder attempts of a giddy rabble . . .',[22] and a wall was built to protect them. By 1775, they were 'ruinous'.[23]

I keep the longer elevation on my right, walking down the narrow way between two long brickwork flanks. After a while, the elevation on my left tapers out, but the one on my right still runs seemingly for miles. Between 1786 and 1791 this new ropery was built to replace its Stuart predecessors. In full, the building extends for 1,140 feet, then the longest building in England, now seven times as long as Nelson's column is tall and a much more vital monument to his age.

Like the ropes spun within, the ropehouse is simple and purposeful. Unlike the debauched trappings of Buckingham's water-gate, the building is dressed in sober, harmonious archi-tecture: its simple brickwork and regular window openings embody the order and proportion which characterised Georgian architecture and which ruled this dockyard. It somehow manages to be more aristocratic in bearing than the water-gate, even though the latter was built for a Duke; the ropery is aristocratic in scale rather than architecture, dwarfing the shrunken reach of the early Stuarts which the water-gate embodies. It may seem unfair to compare a Duke's water-gate with a nation's ropehouse. Yet, as no Stuart naval architecture survives, the water-gate is the next best thing.

And looking at plans of Stuart roperies, it's plain that this building was much more than just a renewal of them. The Chatham Ropery is a double ropehouse, combining the laying floor at ground level and spinning on the floor above. Tar barrels were stored in the undercrofts, while apprentice ropemakers trained in the attics. It was modelled on the more recent Portsmouth double ropehouse, which had been built a few years previously after the ropery there was destroyed by fire. It was by now realised that it would be lax in the extreme to clothe such flammable ensembles in timber, so immense quantities of brick were specified for the walls. This prudence was rewarded at Portsmouth, which in 1776 survived an attempt by an American sympathiser to burn it down; after his trial he was hung from the sixty-four-foot mizzenmast of the frigate *Arethusa*.

Sadly, the names of these roperies' architects have not come down to us. Both Chatham and its Portsmouth prototype were designed by unknown individuals in the Navy Board. The only names we might attach to the building are those of Lord Sandwich, who ordered the rebuilding of the Chatham ropery, and Messrs Samuel Nicholson & Son, who built it. I'm reminded of the nameless individual who invented the ships' wheel, thus freeing vessels from the limits of the whipstaff, and the countless anonymous crews who hauled behind their figureheads. For we must not forget that despite the bewildering variety of a ship's parts, and the wider skein of geopolitics in which the Royal Navy operated, the story of the navy is ultimately the story of the vast efforts of people whose names lie buried in the records or are not recorded at all. Behind every sail there is a profusion of hands.

To walk the length of the ropery is to pace out the fathoms of a ship's cable, to retrace the lost geometries by which ships were worked, lines of battle held, empires formed, trade maintained. Near the end, I come to an opening in the flank reached by a short flight of steps: a way into the laying floor, on which rope still rustles in gestation. I poke my head through the door with the sense that I am trespassing, though no ropemakers materialise to rebuke me. Within, the ropehouse looks like a horizontal mineshaft, ropes and machinery trailing to a vanishing point in the gloom. Or as though the curving innards of a ship of the line had been straightened out and infinitely extended: these long views are timber-framed, the ropehouse interior being floored, ceiled and columned in woodwork. Despite the brick walls, it still looks like a firetrap. Along the floor are laid neat lines of cordage in various stages of formation. Abruptly one of these briefly slithers to life, pushed and pulled by unseen forces, before lying inert again.

The new ropery was the product of an age in which aspects of the navy were refined rather than innovated. Ships of the line, lines of battle, frigates, manning, ropemaking – all these things had been pioneered in earlier centuries but were brought in the eighteenth to a perfect coalescence. So it was with the new double ropehouse. It is a monument to the greatest reach and sophistication of sail. Subsequent advances would see it displaced by steam and iron. The marvel of the eighteenth-century navy, as enshrined at Chatham Dockyard, is that the techniques of sail, and the ships thus driven, had been brought to a pitch of perfection.

Returning ropemakers of Nelson's time would recognise the methods by which they are still made here; the ropemakers still use the first items of machinery introduced to industrialise the process, the oldest being Henry Maudslay's forming machine of 1811. The interior is a living world of timber and cordage, far more alive than HMS *Victory* which, for all its cramped Georgian atmosphere, ceased to live long ago. And it is this which prompts me to recall Nelson's Column. If that stone mast immortalises one man, then the millions of bricks which form the ropery surely speak of the sailors who lay behind the deeds of figureheads and flagships. In this sense, the stone coil of rope beside Nelson is perfectly apt. It points the way to here. I reach the end of the building, the end of the dockyard in fact, my progress arrested by a smooth, insurmountable brick wall. If only someone would throw me a line.

I leave Chatham along the old High Street straggling towards Rochester to the south of the dockyard. This was the original core of the town, now engulfed by urban sprawl. On the way there is a brick complex of almshouses for infirm seamen and shipwrights, founded by John Hawkins in 1592, salving that slaver's conscience in the manner of the day. They were rebuilt in 1789 in their present form. Nearby, now demolished, was the Chatham Parish Workhouse of 1725, considered noteworthy by Edward Hasted for its 'large and extensive plan'.

At parish workhouses like this one, old rope would be reborn. Just as ships' timbers were reused in dockyard architecture, so rope had a useful afterlife. Old ships' ropes were picked apart into tarry fragments of hemp known as oakum, which was then sent back to the dockyard to be hammered into the crevices and gaps between the hull timbers to waterproof or 'caulk' a ship. In this way, rope continued to be vital to a ship's life – in the form of oakum, it kept it from sinking. Oakum was picked in parish workhouses like the one at Chatham, where the poor, the weak and the elderly – perhaps once ropemakers, some of them – blackened and reddened their fingertips on the tarry old ropes.

If the Chatham Ropery is a miraculous survival of a Georgian ropemaking operation, then Chatham town feels like a great industrial building with the machinery removed. It's clear that the dockyard kept the town afloat. Now, there are no signs in the High Street of the plethora of sea-going businesses which would have occupied the narrow frontages. Indeed, successive makeovers have made parts of it look like any English high street. This is a shame, for seagoing was what called

Chatham into being and distinguished it from other places. Nothing comparable to the lost energies or crafts of the dock-yard seems to have refilled Chatham town, which makes the ongoing operation of the ropery seem of huge importance. It keeps under tow an increasingly remote past.

And ropemaking lingers spectrally in naval affairs. A ship's speed is still measured in knots; 'hands' remains the term for individual members of a ship's crew. I look at my own as I walk through Chatham town. They are smooth and un-roughened, much like some of the town's streets which, pe-destrianised and freshly paved, have seen no work and little traffic. In contrast, the worn dockyard buildings and the great ropery were atmospherically calloused with their age and their use. It reminds me of the essential tactility of the sailing navy, its direct appeal to the senses which subsequent technologies lacked. Perhaps the Ropery is not just a monument to the navy, but a monument to the hands.

8. The Ropery interior

Bell

The Lutine Bell

CITY OF LONDON
GEORGIAN AND VICTORIAN

I walk beneath glass buildings which glitter upwards in the sea's hues. Some of them are tinted like the dark depths, while others are as clear as the shallows. Threading below them, I tread the deserted streets towards the steel building lying in this press of glass. It was built in 1986 for Lloyd's of London, an ancient body of insurers. A kind of liveried footman admits me to the building's inner sanctum and leads me to an oaken rostrum in the centre of a modern office. There, a golden ship's bell hangs in silence.

On my way here, I had reflected how my treasure-hunt seemed always to be hauling me back from the coasts towards the capital. Yet it is fitting, for the majority of Britain's overseas trade was for centuries administered from the City of London. Though the landing of wares has long since ceased at its quays, the City still has the tincture of the oceans. Besides the seafaring saints of its hallowed churches, the old stone offices are profuse with carved anchors, lighthouses, prows and ships' wheels, tokens of the pride of the long-lost merchants which built them. The newer buildings reflect the sea unknowingly; their giant sheets of aquamarine glass are like some modernist abstraction of the waves.

A ship's bell is unremarkable on a Georgian quarterdeck, but a fascinating presence in a modern office. Sometimes such relics are more acutely themselves when torn from their maritime context and placed in another. This one, known as the Lutine Bell, was recovered fifty years after *Lutine* sank in a storm of 1799 off the Dutch coast. She had been laden with a colossal quantity of bullion to prop up German banks during a late eighteenth-century economic crisis. Naturally, the treasure attracted numerous salvors. In a series of attempts continually frustrated by the shifting sandbanks in which she lay, fragments were raised of *Lutine* and her cargo, including this bell. It's now a corporate talisman for Lloyd's, the insurers of the cargo who paid out the claim in full.

Lloyd's were the pioneers of marine insurance, super-charging oceangoing trade. In the late seventeenth century, England's burgeoning commercial empire depended entirely on the sea and there was an acute need to offset the risk to merchant vessels from war or weather. At Edward Lloyd's coffee house in Tower Street, merchants, shipowners and sea-captains gathered to pool the risks of one another's ventures. Since a lone merchant could not insure the whole cost of a shipment, the risk was split into pieces, with various merchants writing their names under one another's until the risk was fully 'underwritten'.

After the Lutine Bell was recovered, in 1858, it was hung in Lloyd's Leadenhall Street headquarters, by now a corporate office instead of a coffee house. It sounded the fates of overdue ships – rung twice to indicate a safe arrival in port, tolled once to announce a loss. Insurance-wise, a loss triggers a claim and possibly a payout, with those who have underwritten the fated ship's cargo bound to disburse their respective portions if the facts of the matter are not disputed. Consequently, the bell seems to tremor with the economics of shipwreck – the uncomfortable balance between the value of a life and that of a loss, between that which is quantifiable and that which should be priceless. I get up close to the bell and am about to touch it when, behind me, the liveried attendant coughs respectfully. Hand arrested.

The bell has the quietly disconcerting presence of something that has lain long on the seabed. Aside from some distress to its lip, there are no visible marks of its fathom-time – it gleams just as brightly as it ever did at the hands of *Lutine*'s crew. But it is not the same bell that rang the watches upon that deck, having been altered, visibly and in spirit, by shipwreck and long immersion. Nor is it just a badge of corporate pride. It's now a summons to consider what, over the years, we have chosen to raise and display of our seafaring past, and what we have preferred to leave in the deep.

'About last Easter, we chanced to find in the sea at the Needles 200 blocks of tin, a bell and certain lead which I did handle and see under water, being there perished and forsaken . . .'[1] The twenty-year-old Jacques Francis paused, exhaled. He was a diver, giving evidence in defence of his Venetian employer against a charge of theft brought by some Italian merchants in Southampton. Francis specialised in freediving salvage and had been brought to England to probe the murky seabed for the lost cannon and bullion of an English flagship. No Englishman possessed his talents. No Englishman could then raise sunken objects and very few of this pasty, pugilistic people could even swim.

Francis was very probably the first Black person to have their voice recorded in England as direct speech (albeit through a translator). Born some time in the 1520s, his home was an island off the coast of Guinea (West Africa), his childhood an amphibian one. People on these shores were far ahead of others in their diving and swimming. They dived for cowries from the seabed, gold from riverbeds, valuables from shipwrecks. In *Black Tudors,* historian Miranda Kaufmann compellingly describes how Africans' freediving skills were unique, renowned within Renaissance Europe and the Atlantic World. Only these people had the psychological stamina and breathing control to reach the bottom. And the courage. For Francis to have spent time exploring the seabed of the Solent, in the sixteenth century, he must have been pretty brave. The English and Guinean seabeds cannot be very alike.

In June 1545, setting sail from Southampton against a burgeoning French invasion fleet, the Tudor flagship *Mary Rose* had been heeled by a curt breeze and sunk by a low row of open gunports. She was not far offshore and her sinking was a public spectacle, watched by the King himself. Expensively built, crewed, victualled, ornamented and armed, she had been Henry's pride. Immediately minds moved to the question of what could be salvaged from the wreck. Certainly not the hundreds of men who had drowned, but perhaps the valuable ordnance on board, worth an estimated £2million today, might be recovered to save face. So, they engaged a Venetian salvage team who, in August 1545, made several unsuccessful attempts to lift her, even though she lay in comparatively shallow water with her topmasts breaking the surface, even though the King had loaned for the effort two of his fleet's mainstays: *Samson* and *Jesus of Lubeck*.

It was Jacques Francis who led the first successful recovery of the cannon from the *Mary Rose*. He was part of a team assembled by another Venetian, Piero Paolo Corsi, which in 1547 succeeded in traversing the six fathoms of water and 'taking certain guns out of the ship drowned'.[2] Strikingly, despite the fact that most histories of race in Britain begin with the slave trade, Francis appears to have been a freediver in the literal sense. In his own testimony during Corsi's trial at the High Court of the Admiralty, he referred to himself as Corsi's servant instead of slave. That he was paid for his work bears this out. And, indeed, his rare skills as a diver made him a hard man to replace. Thus, probably the earliest directly recorded words of a Black African in England are not those of a slave, but of a salvor.

For the time being, neither salvage nor slavery were English concerns. There seem to have been slaves in England, but there were no laws authorising slavery and there were no English slave-traders operating formally in Henrician times – the trade was instead largely in the hands of the Spanish and Portuguese, who had established it to maintain the labour force in their American colonies. Jacques Francis's appearance at this Admiralty Court suggests that the authorities were prepared to accord him the same respect as other foreigners. Were you reading English history in order of date ascending, the episode might seem a promising harbinger of equality in Anglo-African relations. Unfortunately, it set no precedent.

*

In the 1520s, the decade of Francis's birth, the *Jesus of Lübeck* had been built in a cold Hanseatic port. After a few decades in the service of the Hanse merchants who had commissioned her, in 1544 she was acquired by Henry VIII. A few years later, hers and Francis's fates intertwined – he succeeding where she had failed to raise salvage from the wreck of the *Mary Rose* – and around the same time she was painted for the splendid 'Anthony Roll' of Tudor ship-portraits. (These were presented to Henry in 1546; I wonder what he thought of the painting of his just-lost *Mary Rose* included in the roll.) It shows *Jesus* as a sprightly-looking thing, festooned with streaming pennants, flags and protruding cannon.

Yet by 1564, when she makes her next foray into the historical record, she was over forty years old and rot-chequered. Her decayed state mirrored the evil purpose to which she was to be put – flying the royal standard on John Hawkins's second slaving voyage. For, just a few decades after Jacques Francis' voice speaks powerfully in the historical record, England would embark on the slave trading that would blight the world for centuries.

With *Jesus* and three other vessels, Hawkins sailed from Plymouth in October 1564. Off north-west Spain, waiting for favourable winds, he famously addressed his crews: 'Serve God daily, love one another, preserve your victuals, beware of fire and keep good company'.[3] Sage, humane advice, you might think. Then he transported over 400 enslaved people from the Sierra Leone coast to Venezuela and then to Rio de la Hacha, Colombia. Although no testimony from these people survives, from what we know of later voyages their conditions would have been awful. When they arrived, trade was terribly brisk. Along with the enslaved people, Hawkins brought wine, flour, biscuits, cloth, linen and clothing. In return, he obtained gold, silver and even advance orders for more enslaved people on a subsequent voyage. He returned to England via Newfoundland, entering Padstow harbour on 20 September 1565. As well as the slave-money, Hawkins carried the sweet potato, tobacco, and favourable reports of Florida as a potential English colony.

Over the coming years, tensions mounted between England and Spain. This was partly due to rising hostilities at sea – of which Hawkins's trespassing in the Spanish New World was a part – but also to the massing of Spain's presence in the Netherlands. In 1567, a new governor, the Duke of Alba, arrived there with a powerful army intended to enforce Catholic strictures. Despite official protestations of friendship, English unease was heightened at this military presence near the cloth markets on which it was still dependent.

After his second voyage, Hawkins had pledged not to return to trade in the Spanish West Indies. But the profits in these voyages were too great to be forgone – and possibly the Queen herself sensed a chance to subvert Spanish activities there and divert them from the Netherlands. So, in 1567, Hawkins once again sailed from Plymouth bound for Africa, after much masquerading to conceal the true purpose of the voyage from the Spanish ambassador. Once again, Hawkins sailed with the Queen's *Jesus* and her approval. For some reason, though, no-one had thought to spend any of the slave-money on repairing the *Jesus*. She was by now virtually a hulk – it was reported later in the voyage that 'in her stern . . . the planks did open and shut with every sea, and the leaks so big as the thickness of a man's arm . . . the living fish did swim upon the ballast'.[4]

This time, over 500 enslaved Africans were taken on the Middle Passage. Arriving in the Spanish West Indies, Hawkins at first traded modestly at Margarita and Borburata. But then the governor of Rio de la Hacha upheld the Spanish King's ban on trading with the English, so Hawkins coerced trade from them at gunpoint. Homeward bound, Hawkins stopped at Spanish-occupied San Juan de Ulúa on the Mexican coast for repairs to his by-now-bedraggled fleet. The next day, the Spanish plate fleet arrived, also homeward bound with the riches of the Peruvian silver mines. At first, an uneasy truce was brokered for convenience, but it didn't last. The Spanish attacked Hawkins's fleet, sinking a number of the smaller ships. From the deck of the *Jesus*, Hawkins directed the English fightback against these superior odds, while Spanish cannon burst holes in her now-derelict hull, shredded her sails and exploded her masts. Using her as a kind of ship-shield, Hawkins and his men unloaded what treasure they could onto the *Minion* and the *Judith*. Finally, they left the harbour and gained the sea, out of range of the Spanish guns, abandoning *Jesus* to the Spanish. Forty-five people were still chained in her hold.

Also abandoned in San Juan de Ulúa was any substantive English participation in the transatlantic slave trade, at least for the next few decades. While this was carried on by Spanish and Portuguese colonisers of the Americas, there was a hiatus in English slave-trading as the more pressing need to arm and fight the Spanish took over. But upon the accession of King James I and the making of peace with Spain, English attentions could once again turn west, to carry on the projects of colonisation pioneered by the Elizabethans and to resume the triangle trade so cruelly begun by Hawkins and the Queen's *Jesus*.

9. *Jesus of Lübeck* from the Anthony Roll of the 1540s. The first known English slave-ship. Depicted here, gaily beribboned, when still a naval vessel

'Hence on the first descent of the Bell, a Pressure begins to be felt on each Ear, which by degrees grows painful, like as if a Quill were forcibly thrust into the Hole of the Ear.'[5] So wrote the astronomer Edmond Halley in the 1690s of the drawbacks of the diving bells then in use. According to Halley, the pressure was caused by the air within these bells compressing as they sank. Variants of the apparatus had existed since classical times: in the fourth century BC, it was described 'letting down a cauldron, for this does not fill with water but retains the air'.[6]

Halley might also have been writing, metaphorically, of the discomfort some feel at hearing of the slaving deeds of our forebears, the prick of uneasiness when the subject is broached, the attenuating pressure as the crimes are enumerated, the eventual, violent refusal to go any further into the depths and the rapid ascent to the safer shallows where these things go undiscussed. I would speculate that this has been the attitude of the British majority in the centuries between the abolition of slavery until the present day. I say this because that precisely has been my reaction when formerly engaging with the subject: like a reluctant diver in a primitive bell.

In April 1691 the frigate *Guynie*, returned from Africa with a cargo of beeswax, ivory, redwood and gold, was making her way from Falmouth to London. In circumstances that remain unclear, sometime around April Fool's Day *Guynie* foundered and sunk in the seas off Pagham, near Chichester. The crew managed to bring ashore the gold, but the rest of the valuable cargo went down with the ship. Her loss might have been just another humdrum case of shipwreck – then a most common occurrence – were it not for the fact that her cargo of ivory was to be salvaged by an astronomer and that she was owned by the Royal African Company (RAC).

By this time, England was an enthusiastic participant in the slave trade. Until 1698, the RAC held a monopoly on the trade and was shipping 5,000 people to the Caribbean colonies every year. Pepys speaks of their ships being 'better manned than ours at far less wages', and further remarks on the 'pretty dinners' the men of the company could give on their slave-money. This was a royal venture, founded by King Charles II and headed by his brother James as Governor of the Company and largest shareholder; it was official English slavery on a scale far surpassing anything the Elizabethans attempted. Even England's currency was shaped by it; from 1663, a new gold coin, nicknamed the Guinea, was placed into circulation using slave-gold like that being carried by *Guynie*. The RAC even coined them. In September 1668, Pepys noted: 'this day also come out first the new five-pieces in gold, coined by the Guiny Company'.[7] Guineas would circulate in England for centuries.

Guineas were coined through the activities of men like Captain Thomas Phillips, who in 1693-4 voyaged to Africa and Barbados in *Hannibal*, a former naval fourth-rate which had been purchased by his patron, Sir Jeffrey Jeffreys, and leased to the RAC. Phillips left a journal of this voyage, a disarmingly candid account of his slave-trading activities. Even across the centuries it makes for grim reading. He writes of the opulence of the slave-castles, the lavish entertainments therein, the brutal heat, the casual cruelty everywhere, the fever-ridden swamps and factories, the torment of the 'musketoes'. Dealing personally with the King of Ouidah who sold them to him, Phillips loaded 700 men, women and children onto the *Hannibal*, each of them branded with the letter 'H' in reference to the ship. 'The negroes', he says, 'are so wilful and loth to leave their own country, that they have often leap'd out of the canoes, boat and ship, into the sea, and kept under water till they were drowned . . .'[8]

It took *Hannibal* two months and eleven days to reach Barbados; time that the enslaved Africans spent crammed into the lower decks, only occasionally let out for exercise. Of the 700 who had been abducted, 320 died on the passage – 'a great detriment to our voyage, the royal *African* company losing ten pounds by every slave that died'.[9] Most of them were claimed by an outbreak of smallpox, while some drowned after jumping overboard: 'we had 12 negroes did wilfully drown themselves . . .'[10] And in a truly sickening piece of self-pity, Phillips whined about the extra work all this created: 'No gold-finders can ensure so much noisome slavery [!] as they who do carry negroes; for those have some respite and satisfaction, but we endure twice the misery; and yet by their mortality our voyages are ruin'd, and we pine and fret our selves to death'. Then, more soberly: 'I deliver'd alive at *Barbados* to the company's factors 372'.[11]

Reading Phillips's journal is a harrowing but necessary task in understanding the sheer evil of the trade which brought forth Pepys's golden guineas. Satisfyingly, he got his comeuppance. As he was conducting his foul work upon the Guinea coast, he tells us, he went deaf in one ear. Voyaging homeward, he was seized with 'violent convulsions in my head', went near-deaf in the other ear and was bedridden until *Hannibal* entered the Western Approaches. It's as though fate had judiciously thrust quills into both his ears.

It wasn't just royals and businessmen who were tempted by the profits. There was overspill into purer realms of scientific endeavour. When *Guynie* went down, the RAC's Deputy Governor was Abraham Hill, also a fellow of the Royal Society. And this is how Edmond Halley came to be involved in salvaging cargo from the *Guynie*. For the RAC, which purchased *Guynie* in 1689, her task would have been not to carry enslaved people but the profits made from them, hence her cargo of gold and ivory, worth the expense of salvage.

A few months after *Guynie* sank, Royal Society minutes record that Halley was relating 'the Success of his Experiments of going under water in his diving bell'.[12] In his later publication, 'The Art of Living Underwater', Halley explained how he dealt with the aural pain caused by the pressure of descent in a bell. It was a simple solution. Accompanying his diving bell, actually a wooden, lead-clad cone, were a couple of air-barrels linked to it with 'a Leathern Trunk or Hose, well liquored with Bees-Wax and Oyl'. This enabled the circulation of fresh air through the submerged bell, which relieved the pressure on the ears and allowed Halley and four others to be 'together at the Bottom, in nine or ten Fathoms Water, for above an Hour and a half at a time, without any sort of ill consequence: and I might have continued there as long as I pleased . . . and by the Glass Window, so much Light was transmitted, that, when the Sea was clear, and especially when the Sun shone, I could see perfectly well to Write or Read, much more to fasten or lay hold on any thing under us . . .'[13]

Bells should speak. On deck, the Lutine Bell made everyone simultaneously aware of the time (shipboard days being sliced, differently to ours, into six 'watches'). Then, for over sixty years, it lay stoppered with sand and water on the sea floor. Later, on the underwriting floor at Lloyd's, its voice was restored: it made everyone simultaneously aware of a loss or a safe arrival. Now, the Lutine Bell has fallen quiet again. Ship losses are announced in less time-honoured ways; news arrives down the wire, rather than through the air; a great crack now runs through this bell, silencing it further. I should like to give it voice again.

Halley says little about the sunken *Guynie*, only of the workings of his improved diving bell and his doings upon the seabed. To him, she was merely another ship, an opportunity for putting his underwater theories into practice. The RAC appears to have held a similar view: she was but a vessel, unworthy of salvage except for her valuables. Now, I think, we should see it differently. For somewhere on the Sussex seabed there lies a ship with tangible links to the slave trade in its Stuart prime; to me, at least, her salvage would be an opportunity to bring us closer to its material culture, to move beyond apologies for the slave trade and into direct confrontation with one of the ships that undertook it.

During his work in the 1690s, Halley gave a number of papers on his diving bell improvements to the Royal Society, which were eventually consolidated into the *Art* and published in 1713. By this time, the RAC had lost its monopoly on the English slave trade, for a number of reasons: the company's debt, the colonies' voracious appetite for enslaved Africans beyond that which any one company could supply, the fall from the English throne of the Stuarts (on the patronage of whom the company depended) and the accession of William and Mary in 1689.

With the English slave trade now free from monopoly, others crowded the field. The RAC continued in a reduced state until 1731, when it abandoned the trade in favour of ivory and gold dust. The same year as Halley's *Art* was published, the Treaty of Utrecht granted the *asiento* to the English South Seas Company. This was the contract, held by a number of European nations, to supply the Spanish colonies with captured people, formalising what John Hawkins had been attempting to do one hundred and fifty years beforehand. Under the terms of the contract, the company would send 4,800 people per year to Spanish America, for thirty years. After frenzied stock-trading, the South Seas Company imploded and the British government surrendered the *asiento* in 1750. That year saw the foundation of yet another English slaving company, the Company of Merchants Trading to Africa. It inherited the RAC's infrastructure, the network of ports and forts including Cape Coast Castle. With these it facilitated British merchants following the abhorrent triangle by now so deeply grooved in the Atlantic and trafficked by ships of all kinds, not just slavers.

'Not being used to the water, I naturally feared that element the first time I saw it, yet nevertheless, could I have got over the nettings, I would have jumped over the side . . .'[14] Olaudah Equiano was writing thirty years after he had been enslaved and taken on the Middle Passage as an eight-year-old. Born sometime around 1745, he was sold and resold between various owners in England and America, eventually buying his freedom in 1766 and becoming an indefatigable campaigner against the slave trade. Published in 1789, his autobiography gave, for the first time, a voice to the millions of African men, women and children who had been forcibly transported across the Atlantic.

His testimony conveys the sheer unreality of the slave ship, which he called 'this hollow place'. Not having had any experience of seafaring, he ascribed many of its aspects to magic, such as the action of the anchor in stopping the ship. 'I was more persuaded than ever that I was in another world, and that every thing about me was magic.' And inhumanity: 'I was now persuaded that I had gotten into a world of bad spirits, and that they were going to kill me.' He described how 'two of my wearied countrymen who were chained together (I was near them at the time), preferring death to such a life of misery, somehow made through the nettings and jumped into the sea'. Another followed, and more would have done if they had not been stopped by the ship's crew, who then stopped the ship and put out the boat to go after those who had jumped overboard; 'two of them had drowned but the third was brought on board alive and flogged . . . unmercifully'. After all, this was profit, not people, that the slavers stood to lose.

Following his arrival in Barbados and transhipment to Virginia, Equiano was sold to Lieutenant Michael Pascal, a Royal Navy officer who renamed him Gustavus Vassa, the latest in a series of names he had been given and to which he first refused to answer, being cuffed until he did so, but then afterwards used throughout his life. (I am not sure whether to call him Vassa or Equiano, but Olaudah was his original name and seems most respectful at this distance in time.)

In 1759, the year of Britain's naval *Annus Mirabilis*, Equiano was baptised at St Margaret's Church, Westminster. By this time, he had spent five years serving as Pascal's slave aboard a series of Royal Navy warships, and had been a participant or witness to several battles of the Seven Years' War. In August that year, he was aboard the *Namur* (she who was found under the Wheelwrights' floor at Chatham), flagship of Admiral Edward Boscawen, as the British fleet chased and engaged the French off the Spanish and Portuguese coasts. In the resulting Battle of Lagos, Equiano's job was to bring powder to the aftermost gun, running the length of the ship to retrieve the powder from the magazine: 'cheering myself with the reflection that there was a time allotted for me to die as well as to be born, I instantly cast off all fear and thought whatever of death . . .'[15]

Equiano's narrative is remarkably free of self-pity and is marked with a generosity of spirit; it is the work of a man at peace with himself despite all that he has endured. His is the first voice of a survivor of the Middle Passage trade to arise from those depths; also, his story shows starkly, uncomfortably, how enslaved people were to be found throughout institutions such as the Royal Navy, which ostensibly were unconcerned with the trade but in fact coexisted with it until abolition. For the Royal navy was not only for defending Britain from seaborne attack; it was also for protecting British trade and harrying that of the enemy. And slavery, trade and war were deeply intertwined.

During the Fourth Anglo-Dutch War, the British privateer *Alert* captured the Dutch slave ship *Zorge* off the African coast with nearly 250 men, women and children aboard. In the practice of the time, *Zorge* was a vessel belonging to the enemy and therefore a fair prize for *Alert*, a merchantman operating under Letters of Marque from the British government which allowed her to prey on enemy vessels. *Zorge* was sold to the Liverpudlian Gregson syndicate, one of scores of such slaving syndicates operating from the ports of London, Bristol and Liverpool. She was renamed *Zong* by her new owners.

At Cape Coast Castle and Anomabu on the Ghanaian coast, *Zong* took aboard more captives to bring the total up to 440, far in excess of the number carried by most vessels of her tonnage. In September 1781 they took on water at St Thomas island and proceeded into the Atlantic. It was a long crossing, helmed by a sick and inexperienced surgeon, Captain Luke Collingwood, short-handed with only a small crew of twenty sailors to work the ship and attend to the enslaved people, a far cry from the 'better manned' RAC ships noted by Pepys.

In mid-November, *Zong* reached the Caribbean and touched at St Kitts for more water. Shortly after proceeding onwards to their final destination of Jamaica, the crew noticed the water barrels were leaking; more disastrously, one of their navigators erred and the ship far overshot Jamaica. By 29 November they were floundering deep in the Caribbean and running low on potable water. To save supplies and prevent a possible revolt among their human cargo (they claimed), the crew decided to force groups of them overboard. Fifty-four people were thrown overboard on 29 November, forty-two on 1 December and a further twenty-six a few days after that: one hundred and twenty-two people murdered in all.

It is this which most starkly reveals how Africans had come to be perceived in the slavers' eyes – as goods, chattels, *things*, not fellow human beings. Over centuries of exploitation they had been deliberately 'othered' by Europeans as sub-human: beasts, brutes, cattle, breeders – carefully honed racism which justified the trade, the plantations and the wealth that flowed from them. And this notion of people as things did not just belong in the slavers' milieu but had been formalised and articulated through insurance law. The people drowned in the Caribbean were underwritten in London.

Zong reached Jamaica shortly before Christmas, where she disembarked her surviving cargo of 208 enslaved men, women and children (a number had died of other causes on the voyage – an additional ten people committed suicide by jumping overboard) at Black River to be sold. Renamed *Richard*, the ship then left for England, arriving in October 1782. The Gregson syndicate subsequently claimed the losses of those murdered at £30 per head, under an insurance contract that covered the death of enslaved Africans on such voyages due to the 'perils of the sea'; the figure was based on the average price attained for people sold at Black River.

Echoing the practices of patrons at Edward Lloyd's coffee house, by this time consortia of merchants and shipowners across the nation insured each other's cargoes. The *Zong* was insured by a consortium of Liverpool merchants headed by Thomas Gilbert. They refused to pay out against the Gregson syndicate's claim for losses (though having insured the enslaved people in the first place, they were not exactly heroes of the piece). Thus arose the infamous court case *Gregson* v. *Gilbert,* tried at the City of London Guildhall in March 1783. There, the jury of City merchants unsurprisingly granted the Gregson syndicate the £3,660 they had claimed, accepting the argument that enslaved Africans were a cargo like any other and could be sacrificed to save a ship.

But the insurers appealed. Gilbert's syndicate disputed the circumstances in which the enslaved men, women and children had been killed – specifically on the questions of water supply and reported rainfall, which would certainly have thrown the claimed low water supply into doubt. In May the case was re-tried at Westminster Hall. In both cases, one of the presiding judges was Lord Mansfield. He was a complex character, a Scot who rose from being the impoverished fourth son of an aristocrat to a position of great legal eminence. In the earlier case of James Somerset (1773), he had ruled that a runaway slave could not be forcibly taken out of England, a ruling widely though incorrectly interpreted as abolishing slavery there.

It happens that my first job was at Kenwood House, Mansfield's former country home. Long after it had been vacated and opened to the public, I was one of the house custodians who stood in the rooms explaining the house to any visitors who cared to ask. We were expected to sell Mansfield's achievements and the gift shop jam in equal measure. We would hail Mansfield's role in the abolition of slavery, but it is only now, a decade hence, that I see that we whitewashed the full complexity of the story. Yes, Mansfield did rule that slavery was illegal in England and called it 'so odious that it must be construed strictly',[16] but his ruling did not suddenly free all slaves in England, as the abolitionists and we, centuries later, had airily supposed.

When the *Zong* case came to trial, among the residents at Kenwood was Dido Belle. She was the illegitimate daughter of Mansfield's nephew, Captain John Lindsay, and Maria Belle, an African slave. Lindsay was a Royal Navy captain who would become the first commander of HMS *Victory* on her commissioning in 1778. He had met Maria while on station in the West Indies. He freed Maria and gave her land in Florida, but left Dido in the care of his uncle. Dido lived with the Mansfields for thirty years, who raised and cared for her and remembered her in their will. Like Jacques Francis, centuries earlier, Dido seems to have held an ambiguous position by the standards of the day, something between family member and servant, though it's certainly clear that she was playmate and later companion for Mansfield's great-niece Lady Elizabeth Murray, who also lived in the house.

Despite Dido Belle's presence in his life, Mansfield was somehow still able to think of the dead in the *Zong* case as property, not people, saying: '(tho' it shocks one very much) the Case of Slaves was the same as if Horses had been thrown over board'.[17] Such callousness was made easier by the fact that this was classified as an insurance trial, even though murder was the elephant in the courtroom. As the trial unfolded, a crucial fact emerged – that it had rained heavily over the ship between the murder of the second and third groups of people, thus calling into question the need to preserve water by murdering the third group. Moreover, the majority of those murdered were women and children – less valuable on the auction-block than men, hardening the suspicion that this was an insurance scam, not an attempt to preserve water or prevent revolt. The fact of the rain led Mansfield to suspend proceedings and order a retrial. He suggested that the insurers might not be liable to pay compensation after all. However, the retrial never took place, probably because the Gregsons had by now realised that they stood to lose.

Like Thomas Phillips in the previous century, the Gregson syndicate – among them two former Lord Mayors of Liverpool – were unafraid of exposing their slaving practices to public view. They pursued their claim through the courts despite the inevitable publicity. Perhaps slavery was still so normalised that they calculated they had little to lose. In this they were wrong, for the *Zong* case galvanised the then-fledgling abolitionist movement into something far greater. Six years later, the Society for the Abolition of the Slave Trade was founded and Olaudah Equiano's narrative published; twenty years after that, Wilberforce's Abolition Bill passed Parliament in 1807; five years after that, the guinea was withdrawn from circulation, an unintentionally symbolic withdrawing of tainted gold from the British economy.

In those landmark court cases, Lord Mansfield seems to have struggled to equalise the competing pressures in his ears. On the one hand, he seems to have been a benevolent guardian to Dido, and his rulings haltingly asserted the rights of enslaved people on English soil. On the other, embedded as he was in the Establishment, he saw keenly the centrality of slavery to the British economy, phrasing his rulings to limit any serious upheaval of the trade. The same compromises attended abolition. Slavery persisted: slavers made fortunes on the now-illegal market; though they ran greater risks, the terms of the legislation were still highly favourable to them. Subsequently, in 1833, the Slavery Abolition Act ended the trade and freed enslaved Africans throughout the British Empire – but only by reclassing them as 'apprentices', so their labour would continue; only by paying their owners £20 million in compensation.[18]

At Kenwood, we would often quote Mansfield as saying in his Somerset judgment that the air of England was too pure for slaves to breathe. Actually, this is a popular misquotation. It seems that no-one said such a thing. Instead, the quote appears to be a collage of different voices at different times, all enunciating a commonly held delusion that England itself was too good for slavery, its air superior to the air of its colonies. And air, of course, was too good for the people crammed into the foetid decks of ships like *Zong*; air was ultimately denied to the women and children who were pushed out through the *Zong*'s portholes and, weighed down by their shackles, sank down through the fathoms.

Air, of course, is essential to the work of salvors. The art of salvage is that of bringing air down to where it does not exist: initially in the capacious lungs of freedivers like Jacques Francis, then subsequently with ingenious devices like Halley's barrels and 'Leathern Trunk, or Hose'. Halley made important advances in bringing fresh air down to his diving bell. But the diving bell, though an effective device for spending time underwater, wasn't adapted for undersea work. It was more of an observatory than a device for freely exploring a wreck. Those within could only study circular patches of the seabed, and to move the bells around was cumbersome, requiring much coordination between those inside and those operating the mast-crane from which they hung.

This history of salvage is, in large part, a history of our priorities. Early salvors descended only to recover goods of immediate financial worth, such as bullion and cannon. In 1687, William Phips managed to successfully locate and salvage a sunken Spanish treasure-ship, *Nuestra Señora de la Concepción*, in the Bahamas. Some £250,000 worth of treasure was raised, triggering a salvage craze. Subsequently, there developed various methods of reaching the sea floor. But it is worth examining the motivations for doing so. Before the system of marine insurance pioneered at Lloyds, a loss was a loss. If cargoes were particularly valuable and lay in shallow water, recovery of the most precious goods might be attempted. But once cargoes began to be insured and risk pooled, loss was offset by compensation, lessening any motivation for the owner to salvage the sunken cargo, even if it were valuable. That same cargo might then be sought out by salvors seeking the profit renounced by the original owners of the goods. Though it was unintentional, the new practice of insurance had harmful consequences for the way we perceive loss. Shipowners now engaged differently with the things in their holds. And this, of course, is what helped to dehumanise enslaved people.

If the act which encapsulates the supreme horror of slavery is the throwing overboard of shackled people, dead or alive, during a voyage, then salvage is the opposite: the bringing up of their memories from the depths. I have followed the histories of the two because there are points where they curiously intertwine - for a different perspective on this phase of history which (thankfully) is beginning to rise to the surface, but mostly because we have a duty to salvage and handle the memory of this abhorrent trade. And, importantly, we are now diving wreck sites not for immediate profit, but for a less tangible and much more important form of reckoning.

Like his father before him, Charles Deane was a caulker, hammering oakum into the crevices between ships' timbers. With his brother John, he had sailed on the ships of the East India Company, profiting from private trade carried on alongside his seaman's duties, but Charles does not seem to have been particularly fond of this life and in 1822 found work in Barnard's shipyard in Deptford. And unlike his more conventional brother, Charles was of an inventive, mercurial disposition. Caulking – ship-saving task though it was – was not enough. While in the shipyard, he fashioned a beaten copper helmet with three thick glass windows, attached to a canvas and leather suit, all of his own devising. Air was supplied to the helmet through a bellows pump, and a second pipe allowed exhaled air to escape.

This was not, as you might suppose, designed as the world's first diving suit. Rather, Charles had intended it for fire-fighting. Patented in November 1823, the apparatus nevertheless soon found use in rescuing sailing ships and their dockyards. That it might be adapted for use as a diving apparatus was suggested by his brother John, who after leaving the East India Company had settled in Whitstable, then a centre of English salvaging. With a few modifications they developed an 'open helmet dress', with the copper helmet mounted on a kind of vest with open joints between it and the rest of the canvas and leather suit. Although the design was ingenious, the suit was severely limited by the danger of water seeping in through the gaps and flooding the helmet if the diver leaned too far forward.

To manufacture their suits, the Deanes went into partnership with the German-born engineer Augustus Siebe. In his Soho workshop, Siebe was able to manufacture all the required components and in the late 1820s was also able to devise a solution to the limitations of the Deanes's 'open helmet' design. The result was Siebe's closed dress. When you think of old-fashioned diving gear, you are visualising the Deanes and Siebe's designs, or refinements of them: beige India rubber suits with gleaming copper helmets. Remembering their first employers, the Deanes first used their new suits to salvage the wreck of the East India Company vessel *Carn Brea Castle*.

The ship had foundered off the Isle of Wight in August 1829, and the insurers had commissioned the Deanes's salvage of its cargo of copper and iron. The East India Company, of course, was not just an innocent trading concern but an extraordinarily powerful commercial force which by this time effectively ruled India for profit. It would continue to do so until the Indian Rebellion of 1857, when it was nationalised and appropriated by the British Government, who carried on British rule on the subcontinent through much of the Company's infrastructure. As with the slave trade, fabulous wealth flowed into Britain from sources unseen, offshore, submerged; in exchange, European culture and innovations were transplanted to these places, of which the true legacy remains the subject of torturous debate today.

The Deanes' salvage of *Carn Brea Castle* was successful and their suits went into widespread use in wreck recovery projects. And in 1858, Siebe's suit was worn by divers mounting yet another attempt to salvage the precious cargo of *Lutine*, after thirty years in which she had lain swaddled in sand. She had originally been a French frigate, *Lutin,* captured at Toulon in 1793 and recommissioned into the Royal Navy. Her name originated in the French 'Lutin', roughly translated as 'sprite' or 'imp'. And there was something impish about the way *Lutine* had eluded the salvors' hands.

After she sank, the Royal Navy guarded the site for a time, endeavouring to raise her: they failed. Then, Dutchmen from the nearby coasts made a series of dives and carried up some gold bars and thousands of Spanish *pistoles,* as well as swords and other artefacts. A later series of attempts with a diving bell failed to reach the wreck. A 1697 description of the folkloric lutin might easily be addressed to the wreck of the frigate: 'you are invisible whenever you please . . . you descend to the centre of the earth without dying . . . you plunge into the depths of the ocean without being drowned . . . at any moment you please, you re-appear in your natural form'.[19]

Then, in 1857, Dutch fishermen happened to find her again. A year later, helmeted divers worked the wreck site, salvaging gold and silver bars, *pistoles*, cannon, the rudder (remade into a chair and table for Lloyd's chairman) and, in July 1858, the ship's bell. It was found tangled in the chains that linked the ship's wheel to the rudder. Curiously, it is engraved not with the name 'Lutin' but 'St Jean 1779'. Right date of launching, but wrong name. A few theories have been advanced as to why, though we will never certainly know. A last-minute name change for the ship, the bell having already been cast? A bell from another captured French prize, of which there were several so named? Or a reference to St John the Baptist, whose protection was sought?[20]

By the time of the bell's recovery, the English slave trade had largely been stamped out by the Royal Navy, which in the years following abolition had formed and dispatched a Preventative Squadron to cruise the slaving-grounds off the West African coast and the Middle Passage. And by the time it had been installed in Lloyd's, slave-ships had ceased to be underwritten by syndicates in Britain.

In 1823, the navy purchased *Black Joke*, an ex-Brazilian slave-ship recommissioned into the service on account of her extraordinary speed. In January 1829, she chased the Spanish slaver *El Almirante* for thirty-one hours, killed most of her crew in battle and freed 466 enslaved people on board. Between 1810 and 1849, such exploits resulted in the Royal Navy freeing 116,000 people, although millions more perished en route to Brazil, last redoubt of the trade. Admirably, the navy's efforts here were not just tokenistic. But there lingered social and cultural legacies of the slave trade, the endemic racism and inequality that arose from the trade and were woven into the nation's fabric. No attempts were made to unpick these, or to seriously reappraise the reputations of those who had conducted the trade. Instead, barely a century later, statues were enthusiastically raised to their memories.

Once, he used to stand in that leafy square, middle-aged, leaning pensively on a stick. In fact, he looked positively haggard: lined face, unseeing eyes fixed upon the floor, jawbone resting heavily in his left palm. This man should not really have had much cause for dismay. On the plinth below his now-absent feet, an inscription calls him 'one of the most virtuous and wise sons' of the city. Perhaps his bearing was supposed to convey that virtue and wisdom accumulated during his long mercantile career. But the effect of his posture was to make him seem uncannily guilty.

Born in Bristol in 1636, Edward Colston was long a celebrated son of that city. After an apprenticeship to the London Mercers' Company, Colston flourished as a merchant, building up a wide web of interests in the Iberian Peninsula, the Mediterranean and Africa. He also held numerous directorships in the Royal African Company, rising to serve as Deputy Governor in 1689-90, even gifting a large chunk of his company holdings to William and Mary after the Glorious Revolution. For much of his career he was London-based, his family having moved there early in young Edward's life; in middle age, he began to remember his native city with a shower of benefactions, founding almshouses, endowing schools, embellishing churches. Some while later, Victorian Bristolians judged him worthy of a statue.

First mooted at an 1893 dinner of civic worthies (the Colston Fraternal Society), the Colston statue had a long gestation. Initial subscriptions raised only paltry sums - £201 the first time, £407 the second. Surprisingly (or not), 'the members of the Colston societies were not too eager in the matter'.[21] Then the appeal went out to Bristolians at large, but still the sums collected fell far short of the cost of a bronze. Nevertheless, a Statue Committee was formed, a sculptor chosen ('the commission was placed in the hands of John Cassidy of Manchester'), a casting foundry found. Finally, a 'Handcrafts

Exhibition' was arranged to scrape together the outstanding cost. Colston was cast in Coalbrookdale and erected at St Augustine's Pleasure Ground. Then arrived the fateful day: 'favoured with regard to the weather, the gathering attained to immense proportions, and the scene in St Augustine's was one which will be long remembered. The pleasure ground and the adjoining roadways were densely packed with people as far as the eye could see . . .'

We are now beginning to speak more widely of nefarious cargoes; we are living through a great salvaging of atrocities. A slow-moving reappraisal of Britain's slaving and colonial history accelerated in 2020 with the Black Lives Matter protests caused by the killing of a Black American, George Floyd, in America by police there. During that febrile summer, protests were mounted in cities across the world.

On 7 June, the pleasure-grounds around Colston were again densely packed with people, though this time civic dignitaries stayed away. There were cheers as Colston was wrenched from his pedestal and dragged to the quayside. As he teetered there, a battery of smartphones fired on him as he dropped beneath the greenish water of the Avon. The felling was a dramatic raising of the city's slaving past, for so long submerged beneath civic whitewash. Said the Mayor of 1896, unveiling Colston: 'I trust that this statue which we shall unveil will be an encouragement to the citizens of to-day to emulate his noble example and to walk in his footsteps (applause).' Said the Mayor of 2020, the day after his immersion: 'I can't and won't pretend the statue of a slave trader in a city I was born and grew up in wasn't an affront to me and people like me.'[22]

Statues are obvious focal points for feeling. But statues of slave-traders have a peculiar potency because there is otherwise so little material culture of slavery in the UK. The shackles, the chains, the cannon, the slave-ships – vanishingly few material relics of the trade survive here, reflecting, of course, the way the trade in its prime lay below the surface of English culture. Instead, metal effigies of those like Colston who sustained it and carried it on have stood unremarkably in the nation's halls, squares and streets. Often erected some centuries after the deaths of those they represent, they are usually instituted though the benefactions of the charities with which they laundered their slave-gold. Many buildings, of course, were funded by the trade – it seeped into the foundations of cities, churches, stately homes and monuments; very little of the architecture was unaffected. Perhaps the most obvious architectural monuments to slavery in the UK are the West India Dock warehouses, completed in 1804 as places to store the goods from the first and third lines of the triangle. But unless explicitly written into the masonry, architecture speaks only indirectly of its founders and funders. Statues speak more directly. And slave-ships more directly still.

Some see the toppling of slaver statues as an attempt to rewrite history. There is certainly a debate to be had about whether removing slavers from pedestals removes slavery from public view. However, history does not exist only in effigy; and few would weep if slavers existed only on paper .* No, there are better ways to remember the trade for what it was, and demonstrate our contrition at what happened, and one of these is to embark on a great act of salvage. Raise the ships like *Guynie* and their artefacts we know to litter our seabed; raise them not for their monetary value but for the profit of atoning for their voyages. On every vacated pedestal, place a brute article of slaving recovered from the deep.

Very few slave shipwrecks have been identified, dived and salvaged in British waters. One tentatively classified as such is the wreck of the *Douro*, lost off the Scilly Isles in 1843. This was a decade after the abolition of slavery throughout the British Empire, but the trade was still being practised clandestinely by many. A Liverpool ship, the *Douro* had aboard a great quantity of manillas – horseshoe-shaped metal artefacts, like bracelets, that were traded for enslaved Africans. Purportedly bound for Oporto, the *Douro* also carried textiles and

---

* From summer 2021, the Colston statue was subsequently displayed in 'salvaged' form in Bristol's M Shed.

munitions – goods which Thomas Phillips had carried to Africa in *Hannibal* centuries beforehand: 'I also carried there on account of the *African* company, muskets . . . *English* carpets . . .'

We know of other possible slave shipwreck sites. There is the *Guynie*, of course, lying somewhere off Pagham on the Sussex seabed. On the north Devonian coast, the wreck of the *London* lies somewhere off Rapparee Cove, Ilfracombe. It is said that the ship was carrying sixty slaves from St Lucia, although it must be admitted that the exact status of these people is under debate. Regardless, it would be a powerful gesture of contrition to embark on a programme of identification and salvage of these and other slave shipwreck sites in British waters. To its credit, Lloyd's has pledged to pay reparations for its role insuring the trade. It also seeks an archivist to salvage from its extensive collections items relating to slavery, so that its own role in the trade might be better understood. But there is perhaps one more thing that could be done. Every time we bring up a slave-ship or its relics from the deep, toll the Lutine Bell twice, for the arrival of a vessel long overdue.

10. The Lutine Bell – photograph of 1915
showing it in Lloyd's previous home in the
Royal Exchange. The bell is being rung by
a liveried attendant

Figurehead

Rosa Tacchini *figurehead*

SCILLY ISLES
GEORGIAN AND VICTORIAN

One by one, the ropes fray and split, and the anchor cable parts. On the seabed, the anchor topples over in slow motion, buried in the cable's sinking coils. On the surface, all is in the wind. Lopsided in the storm, the barque lurches wildly, then rocks northwards. She had fled here from a vengeful south-westerly, but it had run her down. On her prow, a wide-eyed young girl watches the rocks nearing the keel. Islanders call these rocks the Paper Ledges: that is a misnomer. Hull breached, *Rosa Tacchini* grinds aground on the granite. Her figurehead remains expressionless, but does her heartwood stir?

We descend the harbour stairs. Below, the wet walls bristle with limpets. Limpid water laps at the treads, jellylike on the surface, while shelly sand glimmers underwater. Engine throttled, ropes freed from the rungs, our scarlet, wood-boned boat sidles backwards from the harbour stairs. It bobs astern for a moment, then comes about to point northwards, towards a lagoon-like sound. To Valhalla, in conditions far more tranquil than the weather which took its inmates there.

Valhalla is becalmed. No beery hall of warriors, like the Norse legend, this Valhalla is a haul of ships' figureheads held in a loggia. I come through lush foliage to a clearing where the right-angled building seems to open itself like a book, spilling characters. Prince. Lady. Friar. Shepherdess. Turk. Soldier. Damsel. Sailor. Some beasts, too: a dolphin, a lion, a golden eagle chewing a snake.

In Valhalla are held the souls of wrecked ships, if ships' figureheads can be said to hold their souls. Noble, full-size figureheads are hung like captives from each column, while behind, inside the loggia, are a group of smaller, pedestalled figureheads, typically busts or half-sized. When alive, these carvings had several functions: lucky charms, distinguishing marks, foci for feeling. Now parted from their vessels, they resemble still lives: neither dead nor in their element.

Figureheads, of course, were carved to be seen singly, not like this, cheek-by-jowl with their former sisters, comrades or rivals. Each was a fully realised character, carved for the individual gaze, always seen at the head of a much larger hull (another reason why they could never be closely clustered when alive). Grouped together the effect is intense, like a party of extroverts undiluted by the shy, or a novel shorn of bit-parts – except for the occasional stray arm or hand from wrecks unidentified (hauntingly captioned: 'nothing is known about the vessel from which this fragment comes').

If the ropes of a ship are its working sinews, the figurehead represents something more mystical, and serves as a reminder that the ship contends with forces huger than itself. While the ropes embody man's ingenuity in manipulating the wind, the figurehead is carved for luck, appealing for a surety that nothing man-made could ever provide.

Originally, the Valhalla figureheads were kept in the distressed condition in which they were found, giving them the ghostly appearance of effigies, appropriate in view of the lost ships they embodied. Now they are cartoonishly painted. Since their original liveries were lost or seaworn when they were recovered, this decoration is speculative. The false colours seem to amplify the surreal quality they possess when shipless and placed out of water; but something pagan about these figures seems to be revealed by the bright paintwork, a pre-Christian root in luck, superstition and darkness.

The Isles of Scilly are a granite archipelago south-west of Land's End. Five of the islands are inhabited: St Mary's, St Martin's, Tresco, Bryher and St Agnes. Seen from sea level, the view is of low-lying islands in a litter of dangerous reefs and ledges. From above, they have the air of a larger island shredded by the sea.

No-one is quite sure where 'Scilly' comes from, or in

which ancient language the name is rooted. My favourite explanation is that they are named after Scylla, the rock-dwelling monster in the *Odyssey*. Scylla personified a reef in the Strait of Messina; the Isles of Scilly are strewn with them. And perhaps the name came to the isles from Mediterranean vessels in the Classical age, as part of the trade in Cornish tin then flourishing. Homer describes how Odysseus and his crew must navigate between Scylla, the reef, and Charybdis, the whirlpool: 'No crew can boast that they ever have sailed their ship past Scylla without loss . . .'[1] Almost the same could be said of the Scillies in their unreformed days.

And they are probably the most maritime place in Britain. While the coasts of the mainland merely edge the sea, here the sea swallows the isles, horizon ruling every view, the water Venetian in the role it plays. No other island group is quite the same, not even Scotland's western isles. Gull-like clusters of little boats in bays. Water-taxis. Queues on quays. Impassive boatmen. Hulls up in every place: in coves, in fields, in lanes, driveways, alleyways, old droveways, on slipways. Almost all the names of the coves are prefixed 'Porth' – a Celtic word with a sense of entry, and these are Scilly's true doors, not the plane runways.

Further inland are microcosms of England: narrow roads and paths thread pasture and prehistoric landscapes. Small, heavy-looking buildings of cubed granite are like weights dropped to hold the islands down. St Mary's is the largest of them and the de facto capital. Here, Hugh Town, the main settlement, sprawls on an isthmus between two beaches, and the principal fortification, Star Castle, perches on a hill like an aged Elizabethan retainer. On the other islands are minuscule forts, model-like settlements, miniature anchorages.

You might describe their history as staccato, with some periods muted and others markedly eventful. Here, prehistory is uniquely visible: shrines, cairns, menhirs, even whole villages naked and lichening in the air. Their rockwork is a human counterweight to the natural sculpture of their coasts, over which, for millennia, early centuries gently rolled and broke.

The Middle Ages were largely uneventful (but for some comings and goings, notably the planting of an abbey on Tresco as an outpost of Tavistock Abbey). Then, with suddenness, a Tudor and Cromwellian mania for fortification, fearing an invasion (Spanish, Dutch) of the Scillies as a springboard to attack the mainland. Under Edward VI, the building of two poorly placed emplacements: misnamed King Charles's Castle on Tresco (too high above the water to be useful) and Harry's Walls on St Mary's (on a site too small and limited in its arc of fire). Under Elizabeth I, the post-Armada Star Castle, built as an eight-pointed star. Derived from Renaissance Italian designs, the Scillies have no medieval castles of the traditional kind standing to any height, only these occult shapes. During the Civil Wars, the isles were the last redoubt of the Cavaliers before they were expelled by the Roundheads. Cromwell's Castle was afterwards built to supersede 'King Charles' on Tresco, on a better site for its batteries to command the water. And after all this rushing activity, the Scillies sagged back into obscurity for a few centuries.

For most of the time it was a hard place to subsist. There were no forests for building timber, no great acreages of good earth for crops, no natural windbreaks against the gales. Livings on the isles were scratched from fishing or farmland manured with seaweed. Landlords were mixed: some invested a little in schools and quays, while most were largely absentee and left the place alone, their arrearing rents collected by agents living in mean style on St Mary's. Only the sea held plenty, but there were scant means to build and maintain boats and fishing-tackle to exploit its resources. Kelp-burning brought extra revenue for a time, yielding sodium carbonate for glassmaking – but this was soon synthesised on the mainland, and frugality returned.

Yet these were wealthy islands in one sense. Decent health for the islanders was almost guaranteed as a result of a mild, warm climate unequalled anywhere else in Britain. Only the Pest House on St Helen's ever regularly harboured disease, having been built in 1764 as a quarantine station for passing ships so afflicted. Otherwise, Scillonian air is uncommonly pure.

Springtime here is better than a Manchester summertime, while a Scillonian summer rivals those of the Mediterranean. On days like these, the sea itself seems glassy, and it is under these conditions that it is possible to see the half-sunk reefs on which Scillonian well-being partly depended. It seems perverse to praise shipwreck, but those caused by the reefs encircling the Scillies were received by Scillonians as godsends.

On St Agnes, a tranquil cove is named for St Warna, a shadowy Celtic saint about whom little has been substantiated. She is often described as patroness of shipwrecks, although it would perhaps be more precise to see her as the patroness of poor islanders whom wrecks would benefit. Above the cove is a prehistoric well later dedicated to the saint, into which crooked pins were tossed to propitiate Warna and provoke wrecks. But this, like all folklore, should be taken with a pinch of (sea) salt. So far as we know, Scillonians did not purposely practise wrecking. Instead, as a Scillonian clergyman is supposed to have put it: 'We pray, O Lord, not that wrecks should happen / But that, should they happen / Thou wilt guide them into these islands / For the benefit of the poor inhabitants.'

And benefit them they did. First, from the vessel's architecture came timber for furnishings, buildings or boat-building on these unforested islands. Second, depending on the cargo, were salvaged provisions with which to augment a meagre diet, or raw material to process or to trade, or finery with which to embellish the house or the person. There is a surrealism to wrecking. Objects appear according to chance, not choice. You never know what might randomly come ashore and then into the house. Writing in 1794, the Reverend Troutbeck observed: 'you may see them [Scillonian houses] adorned with pictures, ears of corn, and wreck furniture of various kinds; the last of which are sent them by the hand of Providence'.[2] I was told a story of an older woman on one of the islands who remembered the sudden appearance, in her father's cottage, of a quite outsized and beautiful porcelain tea-set (how it survived the wreck, Warna alone knows).

Meanwhile, in the intervals between wrecks, unscathed ships brought custom. Scillonian pilots helped their captains to negotiate these complex waters. Trade with them was possible, too, but that depended on having the means to buy, and most islanders' wealth was limited to the haul of their nets. Most Scillonians, therefore, were either smugglers or receivers of goods such as tea, silks or barrels of brandy, at least until Customs and Excise gained a more secure footing in the islands. And neither were the islanders averse to falling in with pirates.

True, pirates had once been a real menace (as far back as 1209, the records of Tavistock Abbey describe the monks of Tresco beheading a scarcely credible 112 of them in a single day). But latterly, pirates could profit instead of pillage local economies, especially remote and poor coastal ones like the Scillies: they flooded them with stolen cargoes, captured vessels and, most importantly, prize-money. Scillonians also went in for privateering – a form of state-sanctioned piracy against official foes. One of the earliest Scillonian-built vessels, the *Grace* of 1779, captured ten valuable Dutch vessels during her career.

Prior to the nineteenth century, then, these islands lay at a comfortable remove from the mainland, poor in reliable soil but haphazardly rich from illicit seagoing. Life was led in the shadow of obsolete fortifications, to a rhythm of quasi-tropical summers and bleak, howling winters, subsisting on tides and tillage. If luck held.

Superstitions blossom when a people are encompassed by the sea; they grip hard when a failed harvest or an absent shoal would mean starvation. In an otherwise favourable account, Revd. Troutbeck censures Scillonians for having 'several groundless fancies relating to plants . . . many other such whims . . . relicks of the Druid superstition'. Even down to the twentieth century, traces lingered of beliefs shared with sailors: the folly of uttering the name of a land animal aboard, of sailing with a parson, of turning a boat anticlockwise, of whistling in a gale, of changing a vessel's name.

Space for older beliefs was maintained by the remoteness of mainstream religion (despite the presence of Tresco Abbey). Take the reaction to William Borlase's opening of the barrows. In 1752, this fifty-six-year-old antiquarian visited Scilly and spent some pleasant days prodding away at the isles' tombs, cairns, quoits and barrows that seem almost as numerous as the reefs in their waters. Shortly afterwards came a hurricane; the next morning, Borlase met an anguished Scillonian who complained that it had 'almost ruined him and many of his neighbours, that their potatoes and corn were blasted, their grass burnt quite black, and their pease (which in this island is generally good) utterly destroyed'.[3]

Later that day, Borlase borrowed a room in which to shelter while drawing a view of St Mary's. After small talk with the lady of the house, she 'told me that a few days before, they were in hopes of a plentiful crop, paying their rent, and providing meat and cloaths for themselves and children, but that the last night's storm was very outrageous; then asked me whether we had not been digging up the Giant's graves the day before, and smiling with great good humour, as if she forgave our curiosity though she suffered for it, asked, whether I did not think that we had disturbed the Giants; and said that many good people of the Islands were of [the] opinion, that the Giants were offended, and had really raised that storm'.[4] By sunset, this was the common talk of St Mary's; by the next morning, it was the opinion of the archipelago.

To churchmen like Borlase or Troutbeck and their mainland readers, these superstitions were the products of fear, ignorance, credulity or even pagan ('Druid') tendencies. But superstitions arise not necessarily from these, but rather from a desire to offset hardships and to explain reversals which otherwise would seem frighteningly arbitrary: the ruin of crops, the drowning by high sea of Hugh Town and the collapse of cottages wrought by distant earthquakes (Lisbon, 1755). I certainly wouldn't disparage the unnamed Scillonian woman 'smiling with great good humour, as if she forgave our curiosity though she had suffered for it'. No ignorance there, surely, but a complex, worldly grace.

Figureheads, too, sprang from superstitious feeling: that the sea was a lair of gods to be propitiated or was itself a god; that the ship was a being with a soul that had to be expressed somewhere; and most importantly, that the ship could not see unless it had its own eyes. And so early societies drew eyes on their prows, or fixed figurines to the heads of their vessels so that they might find their way. That is all we can really deduce from the ancient evidence, but these founding beliefs in the figurehead resounded with sailors and shipbuilders almost to the present.

Some of the earliest surviving figureheads are Viking, one of which was memorably introduced to the wider public in the first episode of Kenneth Clark's *Civilisation* (1969). To Clark, this awful effigy – savage beak, bulbous eyes – embodied the darkness of the Viking soul (though now it's thought to be earlier in date, the work of an equally fierce proto-Viking tribe). It seems designed to go beyond merely propitiating the sea: it seems carved to put the waves to flight before the ship. With a shiver, I think back to those luckless Dovorians, cowering in their church with these creatures pointed at their shores.

Incidentally, the Viking presence in the Scillies boils down to a single, odd event. Sometime between AD 985 and 988, a host of longships landed at Scilly after an anticlockwise pillage of the British coastline. What happened next is not entirely clear, but allegedly the Scillies escaped harm. Instead, Viking King Olaf Tryggvason met a Scillonian seer, who correctly foretold that he would recover from a near-mortal wound after spending a week lying upon his shield in his longship. Afterwards the seer baptised King Olaf and his followers into Christianity. The story is questionable, perhaps, but further grist to the mill of Scillonian exceptionalism.

In the Norse tongue of Olaf, Valhalla means 'hall of the slain', so it is appropriate that it was the name chosen by Augustus Smith, the Lord Proprietor of the Isles between 1834 and

1872, for the repository of ships' figureheads he had collected from the reefs and beaches of the Scilly archipelago. In this Valhalla, the 'slain' are ships' figureheads, though they came from merchantmen, not warships.

Naval figureheads have a distinct lineage of their own. Snarling English lions were dominant from the Tudor period (one of which is fortuitously preserved outside a pub in Marsham, Norfolk), sometimes embellished with depictions or symbols of royalty. Ostentation came with the Stuarts and lingered on into the eighteenth century until cut back by the Navy Board. First-rate figureheads could be fabulously complex: that for Cromwell's *Naseby* (1655) was 'Oliver on horseback, trampling six nations under foot, a Scot, Irishman, Dutchman, Frenchman, Spaniard, and English, as was easily made out by their several habits . . .';[5] that for HMS *Victory* (1765) was a multifigure allegory of British dominion over the world, with figures of Victory and Peace and mythological beasts below a shield bearing the Union flag and a bust of King George III.

As the Georgian age advanced, figureheads drawn from classical myth became popular, such as those for *Ajax* and *Eurydice*, while others depicted real people: naval heroes such as Lord Rodney or Earl St Vincent. It was as though warships embodied qualities for which Britain preferred to be known: august, cultured, heroic. In 1796, the frugal Navy Board attempted to abolish figureheads altogether, but strength of feeling foiled them. Figureheads *meant* something. For instance, during an embarrassing retreat up the Channel in 1778, pursued by the French, a boatswain aboard *Royal George* blindfolded the figurehead of the late king, so that it might not see the Fleet's disgrace.[6] Nevertheless, naval figureheads began to thin out during the nineteenth century and became gradually debased in quality. Today, survivors are randomly scattered about the former Royal Dockyards or marooned in museums.

Valhalla is entirely different. Here, figureheads from nineteenth-century merchant ships occupy a purpose-built loggia in one man's peculiar vision of the deep. And these mercantile figureheads are more various, more individualistic than the high political, classical or monarchical references of the navy. For merchant shipowners were free to name their vessels however they chose, and adorn them with whatever characters they pleased.

Another difference is that Valhalla was assembled by chance, rather than the quasi-curatorial process which saved the naval figureheads left in dockyards (retained from obsolete warships when they were broken up, rather than being offered up by the sea). For instance, in 1871, the Spanish barque *Primos* struck the Seven Stones reef and quickly sank. Only one of the eleven crew managed to survive, saved by the ship's figurehead. He clung to this wooden maiden until one of the ship's boats bobbed by, enabling him to row to the Scillies. Coming ashore at St Martin's, the maiden was claimed for Valhalla.

In the early nineteenth century, commercial figureheads multiplied as those of the navy began to decline. In manner, they range from skilfully rendered features that begin to approach fine art to a more naive, homespun quality. And both figureheads and their makers held a peculiarly vague status, both in the arts and in shipwrightry: neither quite one thing nor the other.

Ships' carvers, as they were generally known, operated from workshops adjacent to quaysides. Many were illiterate. Many figureheads have the kind of untutored vigour absent in art's more academic branches. Probably the carvers were not particularly wealthy and were unable to travel abroad. Consequently, figureheads were not often carved from life but sprang rather from the imagination, though it was not uncommon for them to incorporate the faces of the shipowner, his family or his sweetheart. Sometimes there is a vernacular surrealism: unusually butch ladies, for instance, or turbaned Turks with English features. Moreover, it is fascinating to think that their work was ultimately sacrificial: artwork designed to last only as long as its ship, yet which would be seen by a greater audience than most fine art, exhibited in as many far-flung ports as the ship could reach.

Perhaps it was this transience that led to the devaluing of

ship carvers as artists. Like their works, they are quite ephemeral in the historical record. Often all that survives to attest their presence are classified advertisements in local papers, such as that for William Baily of Bristol, ships' carver, father of Edward Hodges Baily, RA, who sculpted Nelson for the Column. Indeed, a handful of carvers like Baily begat lineages of sculptors. In 1637, Gerard Christmas and his sons John and Mathias carved the extraordinary embellishments of the *Sovereign of the Seas* for Charles I, including its astonishing figurehead of King Edgar trampling on seven kings. But few rose to these giddy heights of royal patronage.

Despite figureheads being considered an essential part of a vessel, they never seem to have been critiqued or appraised beyond the sailors who valued them:[7] but then the figureheads' use was mystical rather than practical, and to sceptical landsmen they might have seemed little more than superstitious relics. Yet figureheads are a striking branch of vernacular art, uniquely rooted in ancient beliefs. And in posture, attitude and bearing they are unmistakably of the sea.

In Augustus Smith the Scillonians found another sort of figurehead. He was neither an islander nor related in blood or temperament to the previous landlords (the Godolphins since Elizabethan times and, latterly, the Duke of Leeds). While they tended to be aristocratic and remote, Smith was of mercantile Nottinghamshire stock: portraits show a burly, upstanding, clubbable kind of man, Establishment to his core. Outwardly, at least.

After (or, perhaps, despite) an education at Harrow and Oxford, Smith felt a passion for social reforms: education, land tenure, poor relief. Instead of reclining into the leisured existence his family's wealth could have provided, he drove localised reforms of the poor laws in Hertfordshire that would be echoed nationally in 1834. That same year, he opened a 'school of industry', first fruit of his passionate advocacy of education. And that same year, after casting about for other places and projects, he leased the Scillies from the Crown.

Smith soon found that though the isles were idyllic in climate and appearance, they were desperately poor and un-improved; at the time of his arrival they had also recently en-dured a devastating famine. He began to turn them around. Required by the lease to improve the infrastructure, by 1835 he had completed a new church and quay on St Mary's and begun a house for himself on Tresco. Today, this island, cen-tral to the archipelago and commanding its sounds, has a pre-eminent reputation; it is the courtliest isle, well kept to every inch, with a less rugged, more resort-like character than any of the others; Tresco Abbey is where Smith's descendants have lived for nearly two centuries.

And Tresco is where Valhalla lies. First admitted was SS *Thames*, a paddle steamer wrecked in January 1841 on the West-ern Rocks with the loss of sixty-one of its passengers (and only four survivors). Depicting the Thames as a bearded river-god crowned with a laurel wreath, the figurehead was rechristened Neptune by Smith and placed in the garden of his new house. Now, Neptune stands aloof from the loggia later constructed to house the figureheads in the south-west corner. Perhaps it was felt that as the first entrant he should retain his pre-eminence.

By 1848, Smith was established as the Scillies' benevo-lent despot. That year, he published *Thirteen Years' Steward-ship of the Islands of Scilly*, a pamphlet in which he described his major reforms of the islands. These constituted nothing less than a paradigm shift. For instance, under previous landlords, land did not pass to the eldest son of a family, but was divided equally between the children, meaning the farms and estates on the islands progressively shrank until they were too small to support the families which subsisted upon them - which, to Smith, was a 'minute and vexatious subdivision of the soil'. To solve this, he imposed primogeniture. Although unpopular be-cause it usually meant younger boys going to sea or the main-land, this new system greatly increased the productivity of the land.

Equally productive was Smith's system of compulsory education, which he called 'the main lever by which it was sought to raise the social condition of the islanders'. In a move far ahead of the mainland, all Scillonian children were obliged to attend schools, one on each island, funded by a penny rate. This was perhaps his greatest achievement. This schooling ensured that Scillonian children were some of the best-educated in the country; with its emphasis on navigation, it made ships' officers of a considerable number of Scillonian boys, extending the range travelled by Scillonians across the world and adding a formal, technical gloss to their established seafaring traditions.

And he extinguished smuggling; he banned Scillonians from plundering wrecks. The latter move was complex and interpreted with leeway. Unlike smuggling, running a gauntlet of modern laws and duties, wrecking was a far more ancient pursuit not easily renounced. The year of his pamphlet, 1848, was bookended with two instances. On 18 January, the Glasgow schooner *Eagle* was lost on the Western Rocks – though the crew was saved – and its figurehead of a golden eagle chewing a snake entered Valhalla.* Then on 27 December, the barque *Palinurus* was grounded on the Lion Rock. In defiance of Smith's ban, the islanders 'saved' from her fourteen hogsheads and nine quarter-casks of rum, and seventy-one puncheons of spirits; Smith plundered her figurehead of a dashing, sword-wielding count.†

One by one, his collection expanded. In 1851 came a likeness of Tsar Alexander I of Russia, torn from a Venetian brig which had been dragged from its moorings on New Year's Day and mauled on the Mare Ledges. In 1852, a Scottish barque

---

* Officially, this figurehead is unidentified. But the *Eagle* wreck of 1848 seems a very likely source for it, given the name and size of the vessel. Furthermore, in an 1858 letter to Sophia Tower, Smith refers to the relocation of the Neptune figurehead in his garden as follows: *'when Neptune is raised on his throne at the head of the grand staircase . . . the eagle will try to pick a quarrel?'* From this it seems likely that (i) the eagle figurehead was in the collection by this time and (ii) the collection at that time perhaps comprised only a few and perhaps just these pieces, making them more prominent.

† Palinurus is Aeneas's helmsman in Virgil's *Aeneid*.

named *Mary Hay* was holed upon the Steeple Rock and relieved both of her Jamaican cargo of rum and sugar and her figure-head, an elegant lady in evening dress. As if to partner with her, the next year brought the figurehead of a gentleman in evening dress from *Rosherville*, a London brigantine which had sprung a leak, burst into flames, then been beached at Pendrathen as a gutted hulk. For a few years, no identifiable figureheads joined this intriguing party until that of *Bosphorus* in 1861.

Built by William Mumford at St Mary's in 1840, *Bosphorus* was a Scillonian schooner nicknamed the Old Turk, and she traded with the Mediterranean during her career. Her figure-head is of a Turk in a jewelled turban, with a flowing beard and a pistol tucked into the sash around his tasselled robe. In the natural realism of the carving is the sense of a Turk remembered, not invented: quite possibly the carver would have met with such a man on a ship passing the islands.

Prior to Smith's rule, Scillonian shipbuilding was modest, but after his reforms began to bite the industry on St Mary's expanded massively. It flourished during his rule of the isles, no doubt stimulated by his land and education reforms and general urging of the Scillonians into trades and professions. And despite the need to import almost all the raw materials, Scillonian vessels turned out to be surprisingly fine. At its peak, Hugh Town was dwarfed by the ships rising on the slipways at Porthcressa and Town Beach.

Uniquely among the vessels whose figureheads had been salvaged by Smith to that point, *Bosphorus* was never ship-wrecked. In 1861, she was sold to new owners who registered her in Essex. Before she left the Scillies, her figurehead was removed and installed in Valhalla. At face value, this seems a strange move for a seafaring community, counter to the belief invested in the figurehead, surely condemning the vessel to a life of ill-luck. It seems hard to believe that so superstitious a people as the Scillonians would think this a good idea. Perhaps this figurehead held more sentimental value than most; per-haps the ship's future purpose was thought beneath her, excus-ing a change of identity. Or perhaps, gripped as they were by

Smith's social reforms, the Scillonians were beginning to lose their superstitions.

Outside the Turk's Head on St Agnes we drink at the pub at the end of the world. At least, that's what it feels like. St Agnes and the Western Rocks are the south-westernmost elements of this south-westerly archipelago. Between us and the Americas there are only swells.

Before us, ale and pork scratchings on the table, then two old slipways intersecting at the cliff-foot, then the bay of Porth Conger green and limpid, as pretty a view as ever a drinker could wish for. There is more foam on the ale than there is in the wavelets. I think of long congers silently entwining and then parting down there, their cold eyes as fixed and glassy as a figurehead's. And beyond this, the tranquil sea fringed with the other islands: St Mary's, closest, then Bryher, Tresco and St Martin's most distantly of all.

Augustus Smith strove to lift the lot of the islanders, to improve the natural resources of the islands and to oust with the sciences the superstitions that had hitherto ruled them. Some were horrified at his paternal despotism. He was denounced by the philosopher John Stuart Mill, but praised by the historian James Anthony Froude, who considered that he 'possessed exceptional and unusual powers in those islands and had not abused them'.[8] There was naturally great interest in his efforts, for this was a reforming time. The dawn of the Victorian age heralded a surge of interest in 'improvement', a flexing of new technological muscle and an enthusiasm for overturning the status quo. And Smith's efforts to reform the Scillies dovetailed with more epic reforms on the sea itself.

Now that Britain was an imperial power, its archaic maritime architecture and infrastructure could no longer be overlooked. Hitherto, vessels had picked their way through vaguely charted tracts of sea, gingerly avoiding reefs that were often rumoured instead of being precisely fixed and marked. This was bad enough, but navigation was still more perilous by night, when lighthouses picked out the coastlines only weakly and irregularly. Some burned brighter than others, depending on their owners' whims; all shone the same colour, so could be confused. As for the ships themselves, although they were often beautifully designed and constructed, the beneficiaries of centuries of shipbuilding tradition, they lacked the standardised trustworthiness guaranteed by a rule of science established through observation. Seaworthiness and handling qualities could not be predicted of any ship until it was launched, and its punctuality depended wholly on the wind. And as for the chances of a safe passage – well, that depended on luck in all of the above, secured by fidelity to a figurehead and the talent of the crew. And there were few safeguards for them.

Glasses drained, we walk back onto the island, arm-in-arm up armspan-wide lanes, leaning back into brambles to make way for bumbling tractors. The centrepiece of St Agnes is the old lighthouse, a stocky, white-tiered affair of 1680, one of the earliest raised for Trinity House, lighthouse authority for England, Wales and the Channel Isles. Poorly sited, primitively lit and indifferently kept, it embodies the deficiencies of pre-Victorian navigational aids.

While in clear conditions the light was useful enough, it was frequently cloaked by the low-lying fog to which isles like these are prone. And the problem with its position is that St Agnes lies at the end, not the start, of the Western Rocks. Navigators might see the distant light but not the gauntlet to be run to safety. They might see the lone basket or 'cresset' of flame through the miry window panes – but only if it were kept burning, and it often wasn't.

Until 1836, Trinity House was not the sole authority for English lighthouses. Many of them were owned and run by individuals who profited from the tolls paid at port by passing ships. This dual system resulted in the worst of both worlds: profiteering by private operators (by starving both lights and their keepers of fuel) and inertia instead of improvement from Trinity House. But as shipping expanded and shipwrecks multiplied, it became obvious that English navigational aids were inadequate. So, in 1836, the private operators were bought out and all England's lighthouses brought under the sole purview of Trinity House. Then the lights themselves were improved: moved over from naked flames to reflected beams. And then new lights were built in the sea itself.

The obvious place for a Scillonian lighthouse is on the Bishop Rock, a barnacled eminence at the westernmost end of the Western Rocks, which would announce not just these hazards but the presence of the entire archipelago. The first attempt was a stilted cast-iron structure felled before completion; the second was a perfect granite tower.* Augustus Smith keenly followed its progress. In 1856 he wrote: 'The Bishop opened his bright eye on the 1st of September, and is very conspicuous between the two hills of Samson from these windows.'[9] No longer were the Western Rocks hidden to ships approaching Britain from the west. And with the Bishop Rock and its fellow towers, English waters were reillumined.

Raised on reefs that were tiny, remote and barely ever dry, these rock lighthouses engineered the impossible. Symbolically, they signalled a new intent to intervene in a medium previously thought unfathomable. Light now shone in the darkest quarters of the sea, where only fear and ignorance had existed before. And with their construction came the first serious attempts to observe and quantify the sea's behaviour.

While the Bishop was being raised in Scilly, Thomas Stevenson was groping towards a series of wave-laws in Scotland. Jointly with his brother David he was head of the Stevenson family of lighthouse-builders. In the mid-1850s this engineer watched, with quite some patience, wave after wave breaking upon lighthouse towers and harbour walls, measuring them with self-invented instruments. He also collected anecdotal evidence from fishermen, who told him of the wave-patterns they had observed. Afterwards, he was able to tentatively outline a law of the increase of the height of waves in relation

---

* Rock lighthouses have a long and fascinating history, trialled with Henry Winstanley's first Eddystone of 1698 and proved with John Smeaton's third Eddystone of 1759. See my *Seashaken Houses* (Particular Books, 2018).

to their fetch and a formula for the relationship between one wave and another.

Watching waves act upon moving hulls, however, is a far more difficult exercise. Precise observation of a hull in the high sea was then nigh impossible. In any case, a ship is supposed to earn its keep. Voyaging is expensive and hulls afterwards require repair; experiments are uneconomic, in the short term at least. Consequently, shipwrightry for most of history was precedented, rather than progressive. Hull forms passed down the generations – tweaked here, enlarged there, enhanced with better mathematics, but ultimately laid down according to what was known to float. And observations of a hull's handling came not from scientists but from seafarers. This had its drawbacks.

'Even from nautical men it is not easy to obtain statements which can safely be reduced to measure and number,'[10] wrote William Froude, a pivotal hydrodynamicist who established methods of ship-testing and experimentation still in use today. As it happens, he was almost an exact contemporary of Augustus Smith, who was friends with his brother, the historian James Anthony Froude; the circles of these reformers overlapped. Born in 1810, to the Archdeacon of Totnes, William Froude appears to have been close to a genius. He took a First in mathematics from Oriel College, Oxford, and then advanced the geometries of railway tracks and the skew bridges which spanned them. Sensing talent, Isambard Kingdom Brunel had employed him in this work, later commissioning him to study wave-forms and resistance, work which would inform the hull design of Brunel's pioneering steamships.

Froude was able to show how waves, their action and their play upon a hull could be mathematically explained and therefore predicted; he fished his explanations not from the sea itself, like Stevenson, but from a large testing-tank built in 1871 for the Admiralty in Chelston Cross, his Gothic mansion in Torquay. In this body of water – the first ship testing-tank ever built – Froude was able to overturn centuries of accepted wisdom. Hitherto, it had been thought impossible to artificially emulate the sea's motion to inform hull designs. Yet, in his tank, Froude was able to show how the behaviour of model hulls could accurately predict that of their life-size counterparts, ushering in a transformation of shipwrightry, with hulls and their propulsion now formed by extensive testing as well as tradition.

As artificial pools were scaled up to the status of oceans for the purposes of naval architecture, it was as though oceans had been scaled down to mere bodies of water. Shipowners and Sea Lords alike sunk tanks in their shipyards. In 1876, Froude received the gold medal of the Royal Society for 'his researches, both theoretical and experimental, on the Behaviour of Ships, their oscillations, their resistance, and their propulsion'.

Now imperial Britain had illuminated seaways and the ability to construct comfortable new vessels proofed against capsize. But there remained one last cause of shipwreck to be rooted out: the state of the existing merchant fleet and the welfare of the seafarers manning its ships. The existing regulatory system was almost unbelievably light. In 1873, the report *Our Seamen: An Appeal* revealed how shipowners got away with murder. Its author was Samuel Plimsoll, MP for Derby and social reformer. The document is a bundle of outrage. Though haphazard and occasionally inaccurate, Plimsoll's polemic races furiously along, reproductions of insurance documents, bills of lading, diagrams of ships - all with scribbled annotations - stuffed between the pages. Plimsoll at times appears almost unhinged by the abuses he uncovers. But understandably so.

All too frequently, ships left docksides undermanned, unevenly stowed, over-loaded, haphazardly modified or virtually crippled. Astonishingly, the inspection and surveying of merchant ships by Lloyd's Register, the classifying authority, was entirely voluntary. Shipowners were free to send out their vessels knowing they were in a dangerous state of repair or loading; what's worse, the then-contractual arrangements between seafarers and their employers allowed for the imprisonment of any deckhand who refused to embark in an obviously dangerous vessel. And most evilly of all, insurance in this context actually made it more profitable to engineer the shipwreck of an old vessel than its repair and eventual scrapping. Not all shipowners did this, of course, but the existing laws made it possible. To take one example: 'I may have to tell the House [of Commons] of a man, whose name you will hear in any coffee-room or exchange in Yarmouth, Hull, Scarborough, Whitby, Pickering, Blythe, Shields, Newcastle, Sunderland, or in any port on the north-east coast, as one notorious for excessive and habitual overloading, and a reckless disregard for human life, who has lost seven ocean-going steamers, and drowned more than a hundred men, in less than two years . . .'[11] False lights upon coastlines are a red herring. Britain's true wreckers were men like this.

Vested interests in Parliament conspired to wreck Plimsoll's proposed reforms: 'I was in a state of strong excitement . . . I felt utterly alone in my work, and so sick with excitement and fear, that I was compelling myself to think of the poor widows I had seen to keep up my courage . . .'[12] Plimsoll did secure a Royal Commission into the high losses of merchant ships between 1856 and 1872, but it was staffed by landsmen who did not ask the right questions. The government proposed a Merchant Shipping Act to address the deficiencies uncovered, but to Plimsoll it didn't go far enough; he proposed his own Shipping Survey Bill, which was defeated by only three votes. Eventually, in 1876, a new Merchant Shipping Act required owners to mark their foreign-going vessels with a load-line in the shape of a 12-inch circle with an 18-inch horizontal bar; subsequently this was made obligatory for all vessels. This simple mark – subsequently known as the Plimsoll Line – visibly and obviously indicated overloading, but it also had great symbolic significance as a stamp of modernity upon a hull.

With all these reforms, the ways into and around Britain became safer and more sophisticated than ever before; it was a kindred difference to that between a beaten track lit by moonlight, and a metalled road brightened by a gas lamp. And the sense of helplessness which had hitherto governed the seas began to wane. Science commingled with superstitions. Britain's maritime superiority is often defined in naval terms – the most battleships, the best ratings, the furthest reach – but perhaps more significant were the works of reformers such as Plimsoll: the new laws, hulls and lighthouses which transformed oceangoing.

In November 1872, driven from her moorings in the roadstead, *Rosa Tacchini* went aground on the Paper Ledges south of Tresco. She was carrying hooves, hides, wool and tallow from Buenos Aires to Antwerp: just one of thousands of vessels making thousands of humdrum voyages across the oceans. Successive gales broke her on the ledges and her uppermost decks came ashore. Her figurehead entered Valhalla just before the year's turning – the year Valhalla was finished and three months after Augustus Smith had died. It is, safe to say, what he would have wanted.

Folded deeply into a corner of the loggia, *Rosa Tacchini* is now somewhat hidden behind the other figureheads, like a young waif in a jostling mob. Yet she stands foremost among them in one respect. She's the only one in the collection to be kept in an unrestored state. The grain of the girl peeks increasingly through old whitewash, allowing her woodwork to breathe. It makes her companions look suffocated by their new liveries, despite the supernatural vitality they gain by their bright paintwork.

I notice a little spider in the nostril of the lady of the *Primos*. I drift from figure to figure, gravel scrunching underfoot. It's difficult to know how best to engage with this crew, who lean into an absent wind and glower at the bowling green. Their home, this set-square-shaped loggia, feels barely able to contain them. One side of my brain senses this. The other side knows that they are forever robbed of the motion that defined them, whether tearing along over a bow-wave or gently nodding in a harbour. Stillness doesn't become them any more than this fresh paint.

At first, Smith's reforms were resented. But as the changes

11. *Rosa Tacchini* and the *Mary Hay*
(background)

began to bear fruit, Scillonians came to see him favourably. Impromptu parties broke out on St Mary's and St Agnes after he was elected MP for Truro in 1857. Under him, life was better than it had been. Scillonian sailors traversed the globe. Their land bore more. Employment was diverse, incomes better, households improved. Visiting in 1850, the Reverend I. W. North drew a series of favourable comparisons between the islands as he saw them and how they had been when Borlase visited. To cap it all, North observed of the Scillonians: 'they cannot now be charged with the gross superstition which he [Borlase] found prevailing upon them'.[13]

Towards the end of the nineteenth century, figureheads began to disappear entirely. They were harder to incorporate into the iron hulls of the steamships then proliferating. It was as though the reforms enacted upon the sea made their lucky eyes redundant. From his eyrie on Tresco, Smith watched it all unfold: 'It is strange how many steamers now make their appearance here in the roadstead, which proves how far they are superseding sailing vessels' he wrote to Sophia Tower in 1869. These friends' lifelong correspondence provides fascinating glimpses into Smith's private persona. He delights in fair passages between Land's End and the Scillies, weighs into political controversies of the day, swoons over new plants and embellishments for his gardens at Tresco Abbey. The letters also reveal the extent to which the islands had got under the skin of this Nottinghamshire squire. On 1 January 1870, he wrote a gorgeous New Year's Day greeting to Sophia, worth quoting at length:

> Whether your rivers have been frozen and your plains white with snow I have still to learn; but it has snowed here a good deal, so that the islands were as white as a widow's cap, with skies and seas around as black as bombazine and crape. Like those of many a widow, our sorrows were very transient, and our state on Wednesday bore the unmistakable symptoms of migrated affliction, of which on Thursday scarce a

trace remained, nature having relieved herself by hysterics and a flood of tears, and the islands are again as green and radiant as ever. To cheer our widowed state, we are busy looking after various properties to which our late irritations have made us successors, and probate has to be granted and administration to be taken out for a French barque laden with oil mats, which went on shore and to pieces at Porthloo, for an English schooner with timber cast on the sands of Tean, and for a large Prussian barque laden also with timber, hard and fast, with her bottom knocked out, on the rocks under Carn Near. Everything now will be at sixes and sevens for the next six or seven weeks, and the Scillonian mind able to think of nothing but wreck, wreck, wreck . . .[14]

And Smith's attitude towards shipwreck had become complicated. As a social reformer, he naturally encouraged and accommodated the construction of better lighthouses on the islands (though, it must be said, he also became embroiled in various peevish disputes with Trinity House). He did what he could to stamp out the plundering of wrecks. Yet he himself delighted in what the sea providentially cast ashore. In December 1861, describing the slow progress of his building works to Sophia, he blamed St Warna for not furnishing him with materials; earlier that year, in March, the wreck of the *Award* on Gweal, west of Bryher, had provided him fine woodwork with which to panel the staterooms in Tresco Abbey.

And then, of course, there is Valhalla. Hitherto, Smith's figureheads had been placed here and there on the Abbey terrace, but towards the end of his life he decided to make a permanent home for them. He had been contemplating a sea-folly for some time: 'If ever I marry a mermaid I shall be able to build her a most perfect boudoir beneath the glassy cool translucent wave,'[15] he wrote whimsically to Sophia on Bonfire Night, 1870.

Originally a rectangular building open to the east, facing a bowling green, Valhalla takes the form of a deep loggia under a steep tiled roof carried on columns of unchiselled rocks. Smith probably designed the building himself (as he did his Abbey and the new church on St Mary's). Valhalla is a whimsical vision of the deep, built to invoke the lost ships survived by these figureheads. There is something Alpine or, yes, Nordic, in the overhanging, sloping roof form. Beneath, the interior is grotto-like: walled with glimmering pebbles and two tons of shells bought on a whim from German seafarers. Outside, the rocky columns recall the reefs on which the ships shattered and went down. Inside and out, the figureheads set their faces to an absent horizon.

They are smaller than the great torsos saved from warships, and more variable in size, in expression, in quality. If naval figureheads embody the heroic scale of the Royal Navy, in the foreground of Britain's maritime history, then these mercantile figureheads capture something of the ordinary shipping in the background. And while many naval figureheads depict the great heroes of history, in the faces of Valhalla are the features of civilians. For instance, that of the *Rosherville* apparently depicts a wealthy pleasure-ground owner in Gravesend; that of the *Rosa Tacchini*, that small, half-length portrait of a young girl, is thought to be the Italian shipowner's daughter.

Three months before *Rosa Tacchini* arrived here, Smith had left the isles for a routine trip to London. On his way back, he caught a cold, then a lung infection, then succumbed to pneumonia on 31 July at the Duke of Cornwall Hotel in Plymouth. He was greatly mourned by Scillonians, who had come to see him as a sort of patriarch. Although his will provided for the children of several women on Tresco, he had never married, and his lease of the Scillies passed to his nephew, Thomas Dorrien-Smith, who carried on both his benevolent rule and the collection at Valhalla. Of the November 1893 wreck of the *Serica,* his daughter Charlotte later wrote: 'the figurehead is in Valhalla, we went down in the gig *Normandy* and cut it off'.[16]

Stepping back for one final view of them, I realise that they are all archaic characters. Prince. Lady. Friar. Shepherdess. Turk. Soldier. Damsel. Sailor. Characters with ambiguous places in the modern world, their eyes turned towards the past. In this, these figureheads represent more than just their ships: they stand for the seafarers' superstitions which, following the great reforms to the Scillies and the wider sea, were overpainted with science. How curious that they should have been assembled here by a social reformer, an upender of tradition with his eyes turned to the future. Or maybe this was his intention all along. Seen this way, Valhalla looks like a home for fallen superstitions: a hall of the old, unreformed sea.

12. Valhalla, turn-of-the-century
photograph of the unrestored
figureheads

Timbers

Ships' timbers (various)

RAMSCRAIGS, SCOTLAND; LOOE, CORNWALL;
   ROBIN HOOD'S BAY, YORKSHIRE;
   REGENT STREET, LONDON
VICTORIAN AND TWENTIETH CENTURY

There is a myth that I have often come across. Sometimes accurately, but more often fancifully, ships' timbers are said to be present in historic buildings across Britain. Back in Plymouth, I stood at the landing of a staircase said to have been built around a mast from a Spanish galleon. A few years ago, in Ballina, Ireland, I was at a wedding in a castle ornamented with wreckage from Spanish warships cut up on the serrated coasts nearby. And in the Scillies, wreckwood was claimed to adorn many an island house.

It's perhaps easy to see why the idea – the myth – is so widespread. For one thing, there's a certain mystique in wreckwood being used to frame living space; a poignancy in the notion of something roving becoming rooted. In most cases, however, the claim to ships' timbers is simply a way of deepening interest in a building. This might be particularly true of pubs where, after a few pints, the stories begin to wear out of line and warp into the wind. Or perhaps these tall tales emphasise, quite literally, the comfort we find in the stories of ships.

Practically, there are certain points in favour. In past centuries, when the transport network was less elaborate, timber from wrecked or obsolete ships served as convenient building material, especially so in poorly wooded areas. And the similarities between naval and traditional carpentry certainly make it feasible. But proving these claims can be difficult. Many of them float on hearsay. Only dendrochronological testing would verify the date and origin of any given timber. From the aforementioned (dubious) examples I could perhaps take secret chippings to a lab, but I daren't. To confirm the mystique would be to blow it away.

In a few instances, the claim is well documented. Augustus Smith's panelling of his staterooms, for instance. Or the remarkable Chesapeake Mill in Hampshire, which is known to incorporate much of the fabric of USS *Chesapeake,* an American frigate captured after a famous 1813 encounter with HMS *Shannon.* But more intriguingly – to me at least – is the sheer number of claims which cannot be substantiated. The claims that are local in scope, smaller in scale. The undocumented claims which domesticate the sea. Shipbuildings, if I may so term them, are a fascinating tributary of the stories we tell ourselves about the seafaring past.

In satellite view, the sea's mass, always vast, always unknowable, settles at the cliff-foot. Fields buffer it from the tarmac coastal road. At right-angles to the cliff, a long, slipway-like track climbs from the road into the hinterland. Beyond, the terrain is hilly, blooming with gorse, strewn with isolated buildings. In street view, the only inhabited building is what I take to be the low-slung farmhouse, harled in a tone of curdled cream. Zooming closer, the other buildings scattered in this landscape are in fact ruins, now roofless and poor refuges even for livestock. Here in Ramscraigs, on the edge of Caithness, the north-eastern edge of Scotland, only the farmhouse chimney smokes.

I am (digitally) here because of a photograph. This image is inherently beautiful, but what makes it so extraordinary is what it depicts. It was taken in 1971 and shows a close-up view of a ruined croft-house, the roofless interior piled with detritus. The photographer seems to have been squatting in the corner of this rectangular building, for in the background of the photo we see a gable end of naked stonework and one low stone wall. But these are all incidental details. The focal point is a surviving element of the roof structure: a pair of curving timbers which spring from the walls of the building to meet in mid-air. Just below, a smaller piece of curved timber bonds

13. The Ramscraigs croft house

them together and a longer, horizontal beam further strengthens the whole.

When I first came across this photo, I was amazed. For it shows both a fragment of roof structure and a cross-section of an upturned boat's hull. Once the central ribs of the frame of a small boat, the curvaceous forms of those timbers are a response not to the static load of a roof, but the push-pull of the sea. Where they meet, a long keel would originally have been fixed, cleaving foaming currents instead of keeping off the rain.

In Caithness, particularly along this stretch of coast, stone is plentiful but trees are not. And stone, perhaps obviously, is unsuitable for use as roof structure (although thin slates are common roof coverings). It's too heavy and can rarely be shaped into the requisite length to span a building. Only timber suffices, but where to find it on a treeless coastline?

From the Shetlands to the Scillies, this has been a

centuries-old dilemma. In these disparate, sometimes desperate contexts, the reuse of ships' timber in buildings is essential rather than ornamental. It is about fulfilling a basic need for shelter, with no deeper purpose or symbolic intent. Undoubtedly, the builders of the Ramscraigs crofts would have seized any spare timber that would have roofed their dwellings. Decrepit old boats or shipwrecks, with their long, ready-formed timbers, served the purpose very well. There is an architectural charisma created by this happenstance. But, in the case of Ramscraigs, there is hardship too.

Imagine that you lived quite happily in the depths of a glen, far inland, as generations of your family had done since time immemorial. Your house might be modest: a single-storey, turf-roofed, stone or earth-walled cottage; your life might not be the easiest, farming a small patch of ground to feed your family; your landlord and his agents may not be the pleasantest (you have to renew your rental arrangements year-on-year, and they always ask for more). But those things don't really matter, for here, at home, your roots are deeply sunk.

Now imagine that, one year, your landlord abruptly refuses to renew your tenancy, you're relocated overnight to a bleak lift of land overlooking the North Sea, and you're forced into strange, malodorous trades to keep afloat: boat-building, herring-fishing, kelp-gathering. Sure, the landlord has built houses for you and the other families, and the air and scenery here might be healthy, but what is all that when your roots have been torn up?

Over roughly a century, from 1750 to 1850, this was the experience of many families in the Scottish Highlands. The Clearances, as they are now known, were the forced eviction of tenants from fertile inland areas to the barren coastal tracts of a landlord's estate. Often the expulsions were violent. In Sutherland, over-zealous agents beat the resistant tenants and set fire to still-inhabited houses. Once they had been evicted, sheep, reckoned to be more profitable tenants of the land, were loosed to graze amongst the ruins. Out on the coasts, the displaced families lived in cheap new crofts and tried to adapt to a jarringly new way of life.

But such was the scale of the clearances that even the coastal resettlement devised by the landlords soon began to fail: overfishing and overkelping beckoned starvation. In the end, many emigrated, choosing opportunity in the New World over coastal penury in the old. Overloaded, decrepit vessels bore these people to the Americas, or to the seabed. Today, their stories survive in the ruins of croft houses such as those at Ramscraigs, and in a global diaspora of surnames.

Until I learned more about the Clearances, I couldn't understand why there seemed to be so little information about Ramscraigs. It seemed to have no history. Granted, this tiny grouping of crofts is not exactly the sort of settlement to have generated tall tales, but it seemed odd to find nothing at all. Seen against the story of the Clearances, however, Ramscraigs' lack of backstory makes sense. It sprang into being on the whim of a landlord, the crofts new-built for displaced tenant farmers who probably would have emigrated at the first opportunity. It would not have been long before these mean buildings became vacant, then derelict, then roofless, then ruinous.

Up here, the street view might better be called 'track glimpse'. I can get no closer to the croft in question than a distant sight of its corrugated roof. So, I leap again to the satellite, homing in on the blurry shape as best I can, picturing a great chaos of agricultural implements below the hull timbers. Going by the photograph, their width suggests that these probably came from the centre of the boat where the hull is widest. Inverted, in building parlance, they became rafters for this house; the upper, curved timber joining them became a collar, the lower straight timber a tie-beam. It looks as though they required little modification for the new purpose, beyond turning them upside down and mounting them in the walls.

I try to picture the vessel they came from: probably not much bigger than a ferry or rowing-boat, clinker-built with the timbers overlapping in the old style. Given her size, she can't have been used for much more than inshore fishing, angling for subsistence in the unforgiving tides of the North Sea, always within sight of land and yet deeply removed. She must have been one of many of her nameless kind along this shore: scratchily built, hardily used, ephemeral as sea spray.

Anchored in the wider narrative of the Clearances, there seems a deeper resonance in the reuse of these hull timbers. During the evictions, it was commonplace to burn the wooden rafters of the tenants' cottages, so that these family homes would become uninhabitable stone shells. It must have been a cruel irony for the uprooted family that first occupied this cottage to see the boat's timbers upturned above their heads. Not only did their new way of life now hang over them; here was a reminder of just how thoroughly their boats had been burned.

Somewhere unseen, flame rouges the morning fog. Below deck, scrums of scorched gunners haul back the gun-carriages. The barrels are swabbed out, then rammed with powder, wadding and shot. Lastly, finer powder primes them, then slow-matches find the touch-holes. The balls bang from the muzzles, whirr across wave, thwack mainmast (lucky), strike hull (more likely), or mangle face (especially unlucky). Below, the gunners rinse the barrels and repeat the cannonade. On deck, the officers resist the urge to duck.

From *navis*, the Latin word for 'ship', navy is derived; from it also springs nave, the term for the body of a church. And ships and churches share more than this. They fly before winds actual or spiritual. With approximate built forms – tower and body echo mast and hull – they have been analogised from the moment they began to be built. And in no other building is the concept of a destination acuter, or occupants so equated to passengers.

I've come to Cornwall, to a church that has known cannonfire. Returning here, my memories swell to great prisms from the grains they had become. Lancets of light in the walls. The Technicolor weft of knitted, knee-pressed cushion covers. Pungent flowers on the pew-ends. During sermons, a background hum of impatience. But today is overcast and the church is dark. I thread my way around the great timber columns and braces which divide the aisle from the nave. They have a seamanly heft in the gloom. Over the chancel is a run of curiously curved rafters. I have to squint to study them until a kindly old church-watcher, arising from some unseen perch, flips on the electricity.

The church of St Nicholas, west Looe, is like a small vessel moored at the quayside. Once just a medieval chapel, the church stands on the harbour shoulder as it kinks around towards the sea. In views from east Looe, across the water, it seems to belong not with the neighbouring buildings but with the ferries congregating below. It has the simplicity of those small craft: unadorned rubble walls pierced with thin lancets, a diminutive tower not much higher than a mainmast.

Looe is that rare thing in England: a Siamese town, formed of two conjoined individuals. From the foothills of Bodmin Moor, two rivers loop and score through fertile lowlands until they merge in a narrow estuary, either side of which the twin settlements of east and west Looe (named from the rivers) simultaneously emerged in the medieval period. Joined by their dependency upon the same harbour, from circa 1400 a bridge marched in emphasis between them.

Then, as today, their personalities are quite different. East Looe is the livelier, more commercial quay from which was once conducted Mediterranean trade and, later, the export of minerals dug out of the moorland at the rising of its riverine namesake. Variously, the harbour has handled claret, salt, tar, pilchards, copper, granite, arsenic and tourists. West Looe, on the other hand, has always been the quieter sibling, a harbour for ferries and smaller fishing vessels. From the outset, the two halves of the town belonged to different parishes, built their own chapels, guildhalls and other facilities and, in Elizabethan times, were individually incorporated as parliamentary boroughs, each returning two MPs to Parliament. Once, this high degree of parliamentary representation seemed sensibly to reflect their prosperity. But by 1814, when the antiquarian

Daniel Lysons described west Looe as a 'decayed market-town', the Looes were among the rottenest boroughs in England and ripe for reform.*

I've known the place all my life. Uphill from west Looe quay is Hannafore, where my father grew up and where my grandparents lived much of their lives. Often, on holidays there, we would spend Sunday mornings at St Nicholas's. I have a handful of impressions from those fidgety childhood attendances, like the Christmas time an elderly organist mangled a hymn over which, characteristically, my grandmother beautifully carried the tune.

At first, the small fishing community in west Looe seemed only to warrant a chapel-of-ease, rather than a fully fledged church. Accordingly, St Nicholas – dedicated to the seafarers' saint – was built as a daughter chapel to the mother church of the parish over the hills at Talland. After the Reformation it ceased to be used for any form of worship and was instead, in the disdainful words of an 1852 *Royal Cornwall Gazette* report: 'frightfully desecrated, being used as town-hall, a room for banquets, dances, and revelry of every description, a theatre for strolling players, and occasionally, of late years, as a temporary conventicle'.[1]

Revival of its fortunes came in the mid-nineteenth century, when mineral transhipment breathed new life into the town. In 1852, amidst a flurry of civic improvements, St Nicholas's was reclaimed for worship. 'To wipe away this stain . . . by restoring this once consecrated building to its holy uses,' the *Gazette* continued, 'the chapel has been entirely reseated, and otherwise restored, by country tradesmen working under the eye of the Incumbent, after the designs of Capt. C. Cocks.' Among these improvements, the most significant were the extension of the chancel and the addition to the nave of a north aisle on the site of an old gaol cell.

Today, the exterior speaks powerfully of its medieval origins. Internally, however, the nineteenth century has the upper

* They were abolished as boroughs by the Reform Act of 1832.

hand. And, true to the building's form as a harbour church, the interior woodwork is not all that it seems. At the nave's junctions with the lengthened chancel and the new aisle are timbers quite out of kilter with their neighbouring counterparts. Running my eye over these, I wonder how I never noticed them as a child. For their scale isn't that of the church, but of another kind of vessel entirely.

In 1849, a few years before the restoration of St Nicholas's, an old warship was broken up at Devonport, the Royal Navy dockyard in Plymouth. HMS *San Josef* had had an eventful life. Launched as *San Jose* from Ferrol, Galicia, in 1783, she was originally one of the leading ships of the Spanish fleet. However, she would spend most of her career as a British first-rate ship of the line, having been captured by Nelson during the Battle of Cape St Vincent, fought on Valentine's Day, 1797.

That February morning a British squadron, blockading Spain after Britain was ousted from the Mediterranean by a Franco-Spanish alliance, intercepted a Spanish fleet that was about to convoy a shipment of mercury from Cartagena to Cadiz. Not such a fine prize in and of itself, but the mercury was for minting coinage that would support an invasion of Britain. Commanded by Admiral Sir John Jervis, the British squadron of fifteen ships faced twenty-two of the Spanish. Early in the morning they met off the Cape of St Vincent, Portugal; the British formed a taut line, while the Spanish cumbersomely manoeuvred into a pair of straggling herds with a sizable gap in the middle. Through this passed the British line, firing all the while, then tacking around in a 'U' turn to attack the larger group of Spanish ships.

In the age of sail, gunnery drill carried the day. The Royal Navy dominated the seas for so long, in part at least because its gunners were able to maintain a metronomic rate of fire, pouring broadside after broadside into the flanks of enemy ships. There are many instances of British gunnery defeating opponents much greater in size and number. And during the battle of Cape St Vincent, Commodore Horatio Nelson engaged, alone, the vanguard of the Spanish fleet and the monstrous 112-gun flagship *Santissima Trinidad* (*Nuestra Señora de Santisima Trinidad* was her full name). His ship *Captain* was at the rear of the British line and thus potentially one of the last into the fight, so he decided to veer out of formation and cut directly across to the Spanish vanguard. Although theoretically outgunned, *Captain* could deliver far faster broadsides and Nelson held his own until help arrived.

In the melee, two of the Spanish battleships, *San Nicholas* and *San Jose*, stuck fast together. Nelson's *Captain*, by now shot to pieces after hours of close fighting, then rammed *San Nicholas* in the stern. Nelson and his men ran up their bowsprit, swarmed through *San Nicholas* and from her boarded *San Josef*. It was here, after some conclusive hand-to-hand fighting, that Nelson received the surrender of both ships and the swords of their Spanish officers. These he 'gave to William Fearney, one of my bargemen, who put them, with the greatest sang-froid, under his arm'. Since 1513, no officer of Nelson's rank had boarded an enemy ship during battle. And no one had audaciously used one enemy vessel as a stepping stone to another. It was a foretaste of Nelson's stellar unconventionality.

Once the action had ceased, the Spanish survivors fled the Cape, while the British counted their dead and their prize-money. Jervis was ennobled, Nelson knighted. Four ships had been captured and were towed home to England for refitting and recommissioning into the navy. Among them were *San Nicholas* and *San Jose*. The former was adapted for use as a prison ship, while the latter, recast as HMS *San Josef*, became a Royal Navy flagship mounting 114 guns. In 1837, she was retired from active service to become a guard-ship and a gunnery school at Devonport. By 1849, she had lived a long life for a first-rate warship. Worn out, leaking and facing the burgeoning age of steam power, she was finally condemned. We must imagine a scene of similar pathos to J. M. W. Turner's painting *The Fighting Temeraire*, which depicts a Nelsonic vessel of comparable pomp being towed for scrapping by an impertinent steam-tug.

*

All that remains of HMS *San Josef* are the curvaceous ribs joining the chancel to the nave and, probably but not certainly, the stout timber columns and braces between it and the aisle: warship and worship spliced.

Somehow her timbers had caught the eye of the young army captain, Charles Lygon Somers Cocks, who designed the alterations to St Nicholas's. On the face of it, Cocks seems an odd man to choose as restorer of the church. By 1852, when he prepared the scheme, he was a thirty-one-year-old lieutenant in the Coldstream Guards. But although he then had no building projects to his credit, his aesthetic sensibility would later be demonstrated by his involvement in the design of his own house and, in the late 1870s, the planning of Truro Cathedral. And he was well connected. He came from a wealthy family with a string of ties to the Cornish gentry, notably the Bullers of Morval, the Pole Carews of Antony and the Edgcumbes, who overlooked Devonport Dockyard from their estate across the Sound.

Warships the size of *San Josef* yielded vast quantities of good wood. Although most curving, bespoke naval architecture would have had only limited applications, much would be usable: the straight beams and timbers of internal deck-structure could easily be incorporated into buildings. And although we cannot say for certain how Cocks came by the timbers, there are overlapping rings of causality: family ties with the local gentry, his own military connections and his penchant, as an amateur photographer, for loitering in Devonport dockyard to capture picturesque scenes.

But what lay behind his decision to implant wood from the *San Josef* into the church? Possibly it was a matter of practical recycling: here were timbers strong enough for the job, thriftily sourced to keep down costs. Interestingly, the 1852 report in the *Royal Cornwall Gazette* makes no mention of the church incorporating these ship's timbers, suggesting that at the time this was a common practice, unworthy of special comment. But I suspect that there was more to it than mere practicality. Inside, the timbers are very prominently placed, marching amongst the pews, with the curved braces in particular visible from every angle.

It might have been that Cocks felt they marked the divide between the ancient and new parts of the church with a pleasing symbolism. He may have been inspired by the aforementioned synergy between ships and churches, the delight in incorporating the remains of one vessel into another. And then there is the status of these timbers as Nelsonic relics, trophies of the victory at Cape St Vincent and of all that renowned Admiral's conquests; they are positioned in the church as if displayed. But my own feeling is that Captain Cocks came at this from the point of view of a soldier.

The striking thing about such visceral engagements as the Battle of Cape St Vincent is how abstract they were to the nations which celebrated them. Unlike land battles, where there was a remote chance that one might actually see the fighting, even if only from faraway, naval battles fought out at sea could not be gawped at by passers-by. Sealed off by the sea, they were private to the combatants. They were fought in more dimensions than land wars: to the simple principle of shooting at the enemy until they ceased fire was added the difficulty of doing so from a cramped and tilting vehicle, of maintaining a fixed position on restlessly moving terrain, of simultaneously firing while manoeuvring to win the weather gage.*

Only the aftermath was visible on dry land. When they had limped home, crippled ships and wounded sailors conveyed, viscerally enough, something of the *consequences* of seafights. But, for the ordinary person, the near-transcendental collision of wave, weather and weaponry could never be known. Nations praised or lamented what they could not see and what they could barely understand, that which existed elsewhere and would never make landfall.

By 1852, Cocks held the rank of Lieutenant and had served in India and the Middle East. During the Napoleonic Wars, his regiment, the Coldstream Guards, had fought with

---

* A seafaring term for having the wind in your favour.

distinction. Regimental traditions are such that he could not have failed to be aware of these past glories. And he would have been attuned to the disconnect between civilians and service-men, the detachment of nations from the wars fought in their name, and the public incomprehension of the horrors and hardships of battle. By incorporating a warship's timbers into a place of worship, Cocks placed relics of battle at the heart of the community. It was another form of naval engagement.

Two years after his work at St Nicholas's, Cocks was promoted to captain and sailed for the Crimea. This was his generation's Napoleonic Wars; Cocks fought with distinction at the Siege of Sevastopol which, like Cape St Vincent, was glorified by a nation remote from the muddy, pox-ridden grind of the action.

Upon returning home a lieutenant colonel and distinguished Crimea veteran, Cocks had his photograph taken and built for himself, not far from Looe, a house called Treverbyn Vean – a Gothic revival design by the eminent architects George Gilbert Scott (Midland Grand Hotel at St Pancras, London) with interiors by William Burges (Cardiff Castle).The photograph shows a slim army officer in dress uniform with a vast beard and a gentle face. And the house shows the same fascination with ships' timber as he had earlier displayed at St Nicholas, confirming his deliberate intent there. For out of teakwood from RMS *Orinoco,* the ship on which he had sailed to the Crimean War, he built a dining room.

To encounter the proudly displayed timbers of an august warship is one thing; to find those of a humbler vessel in a building is quite another. 'There was this house, up on a cliff, you don't see it very easily . . .' says Stephen, a retired teacher with a knack for making things. A small, cliff-edge cottage at the edge of Robin Hood's Bay, a village crammed into a cleft in the Yorkshire coast. Not a harbour, as such, just a village focused expectantly on a lone slipway down to the shore. Late in 2002, Stephen and his wife bought this cottage as a place to escape York, where she painted and he taught Design and Technology. 'A couple had gone in before us and walked straight out', he says, 'and my wife said she knew then that we'd get this house. It was in a bit of a state, you see . . .'

Having viewed a few others that were a little too expensive or made over with a phoney rusticity, this cottage's state of dereliction suited them well. Built in the early nineteenth century, it had been extended upwards by two storeys in the 1850s and in the 1970s had undergone a brutal renovation, of the sort commonplace in that era. Many historic features such as fireplaces and chimneybreasts had been over-boarded, while the structure of the house had been clumsily repaired. In Stephen and his wife, the cottage found its saviours.

So far, so ordinary, but this was no commonplace fixer-upper. For when he took down the sagging kitchen ceiling, Stephen found, not a series of normal ceiling joists supporting the floor above, but a row of oversized, rough-hewn beams running above him. 'At first I didn't realise what they were . . .'

As his recollections multiply, Stephen's words begin to run away from him in half-formed sentences, crisscrossing like wavelets on a sandy shore. 'Yes, well, you see, I only twigged what was going on when I found those two nameboards: anyone else looking at the other timbers, it's obvious they've come from a boat: if you look hard enough it's obvious, they've got tar and sand impregnated . . . and it was only seeing those . . .' One beam in particular had stood apart from the others. Upon uncovering it Stephen had paused, scratched his head, and phoned his wife. 'I said to her, it's got something carved on it.'

Though described in few words, the scene is vivid: a dishevelled, low-ceilinged cottage kitchen, sunbeams aslant through the windows, a middle-aged couple in their old clothes looking bemusedly upwards. For there, amongst these strange, rough-hewn timbers supporting the floor above, was a newly cleaned length of beam from which gleamed a name in gold lettering, still crisp and bright after 150 years: *Elizabeth Jane*.

Freighted with Sunderland coal, she would plod up and down England's east coast, in at one harbour, out at another, a ceaseless plying of middling ports, for middling cargoes, at middling distance from the shore. Since the 1820s she had served as a coal-basket, lugging black diamonds, as they were once termed, from the coalfields of Sunderland to Suffolk and occasionally London. It was on just such an errand that she was lost. After getting caught in a gale in January 1854, she had been unenthusiastically repaired (her owner had, ten years earlier, been charged with the deliberate sinking of another of his vessels). By July of that year, she was at sea again, heading to Ipswich with another load of coal.

Not long into the journey, off the Yorkshire coast, her crew abandoned her. The conditions were relatively benign, the windspeed reaching only Force 5, but her pumps were choked and she was leaking profoundly enough for the crew to lose faith in her. Unmanned, she drifted melancholically for a while before coming to shore at Ravenscar, a little way south of Robin Hood's Bay.

Two years later, an advertisement appeared in the *Whitby Gazette*:

> Parties visiting the romantic scenery of Robin Hood's
> Bay can be accommodated with respectable Lodgings
> at **Mr. M. Bell's, Belmont Place**.[2]

'Belmont' derives from Old French and means, loosely, 'fair hill'. Moses Bell's lodgings were aptly named, being situated on the cliff-edge high above Robin Hood's Bay, one of only a few houses with sweeping views of the sea. Originally, it had been a small cottage at the end of a track, but in the mid-nineteenth century it became incorporated into the end of Bloomswell, a terrace of small houses. Here, Moses Bell saw an advantage: above his cottage now loomed, two storeys higher, the bare brick gable of the terrace end. It offered him one of the four walls he would need to extend his cottage upwards. All that remained was to procure the rest.

Not much is known of Moses Bell. 'I found him in the census records, I don't know what happened to him in the end,' says Stephen, 'nor what he did, nor whether he has any descendants . . . someone said he was a cobbler, he lived in

Scarborough later on.' Apart from the original advertisement, he appears in the newspapers only twice more: once in a notice advertising a dubious-sounding 'healer', a Mrs Lamb, operating from his property, the other, his death-notice. Says Stephen, whimsically: 'any spiders we saw around the place we used to call Moses Bell . . . good name for a big spider'.

Yet Bell was a resourceful fellow. On hearing of the foundering of the *Elizabeth Jane*, he saw a way to complete his cottage extension. Accordingly, he built the structure of his second and third floors with her deck timbers and nameboards, concealing them behind plaster once completed. Unlike its present owners, Moses Bell was at pains to disguise this visceral intrusion into his house. And anyway, their enclosure within plaster had other advantages, providing thermal and acoustic insulation between floors.

His small ad says a lot about the attitudes of both his time and ours. It emphasises the 'romantic scenery . . . respectable Lodgings' on offer. It doesn't mention the *Elizabeth Jane*'s timbers (though probably he had to pay by the word) because their presence would not have been considered a selling point. Ships were commonplace, even objectionable things to some. Prospective lodgers would not have wished to be confronted with *Elizabeth Jane*'s tarry bones in the cleanliness of their lodging house. They wanted a picturesque ideal of the sea, not its gritty ordinariness.

Unbeknownst to Moses Bell, his small ad was a harbinger. For it signalled the point at which tourists began to trickle into Robin Hood's Bay, soon swelling to a strong tide, eventually displacing all. In the mid-Victorian period, such formerly insular coastal settlements began to receive visitors from inland, in pursuit of more abstract interests: taking the air, admiring the vistas, relaxing by the sea. As this trend gathered momentum, the coastlines were reshaped: along them, ragged contours were smoothed into promenades, while new piers jutted out from them into the sea. Late Victorian England was a golden age for seaside holidaymaking, and many harbours profited from this new traffic.

In so doing, coastal towns sold what attributes they had –
invigorating air, fine scenery – and, eventually, their souls.
Their vitality ebbed and they became performances of themselves. The money to be made from lodging, feeding and enter-
taining visitors would lead to the demise of the more rooted
industries on which they had traditionally depended. At Robin
Hood's Bay, fishing and smuggling were diminished to memories. Now, many of these places are utterly dependent on tourism or are sad wrecks of what they once were.

Times have changed in other ways, too. Writing that
small ad today, we would certainly mention the presence of the
Elizabeth Jane. Unremarkable brig as she may have been, her
timbers would now be a talking-point. Inspired by the discovery, Stephen has researched her extensively, and it transpires
that her working life was not solely that of a humdrum collier.
For she hailed from Guysborough, Nova Scotia, where in 1817
she was built out of fine Canadian softwood on the shores of
a tranquil bay. At just over seventy-six feet long and just over
twenty-two feet wide, she was a medium-sized all-rounder,
designed to carry assorted cargoes to assorted ports. Her first
voyage was to Kingston, Jamaica. Two years on, she appeared
in Hamburg, where she was sold to a merchant from Hull. She
traded with Lisbon, Madeira and Seville. By 1821, London was
her home port; two years later, she was bound for Sierra Leone,
to which she carried unknown goods and from which she perhaps returned with log-stacks of African teak, puncheons of
palm oil, boxes of elephant's teeth and scrivelloes (tusks), sacks
of gold dust or bundles of hides.

Although he tries to play down the significance of his discovery, I sense an undercurrent of fascination when Stephen
speaks about the beams. I ask him whether finding them altered his and his wife's perceptions of the cottage. 'It did, it did
really . . . finding those, they brought it to life, especially when
we traced the vessel back, finding that she'd been to Africa . . . I
mean, we were very taken by it, we did get very enthusiastic . . .
as we found out new things we would tell people and you could
see their eyes just glazing over . . .' We laugh.

I share his excitement. Ships' timbers extend a building's experience (and perhaps, by dovetailing the ages of each, its lifespan). Few bedrooms and bathrooms are as well travelled as those in Stephen's cottage, built out of beams that had weathered Biscay gales, been sunbeaten off the West African coast and braved the frigid Baltic. And, recycled into buildings, ships' timbers place the sea where it is perhaps least likely to be found. Later, Stephen sends me a charming photograph. It's of *Elizabeth Jane*'s second nameboard, inscribed with her home port of Ipswich, proudly displayed alongside his bath, repurposed as a shampoo shelf.

As the Ramscraigs croft had illustrated, a hull's structure, up-turned, bears similarities to that of a roof; a hold's beams and boards, like those of *Elizabeth Jane*, could easily be those of rooms. Of course, there were some differences: the ships' timbers could be oddly notched for demountable screens or weaponry, for instance, or if they came from a larger class of vessel then they were likely to be too big for land use. Nevertheless, housebuilders and boatbuilders walked in the same forests, appraised the same trees, split the same oaks.

The similarity of timber ships and traditional buildings seems to emphasise how deeply the sea once ran in the grain of the nation. But as the nineteenth century advanced, and ships began to be built from ore rather than oak, they diverged from terrestrial architecture; not long afterwards, the sea would start to become abstracted from everyday life. In 1755, Samuel Johnson defined a ship simply as a hollow building with sails. By 1990, in the eighth edition of the *Oxford English Dictionary* (my grandmother's, which I chanced to have to hand) a ship could not be defined in relation to a building, instead being rather indifferently categorised as 'any large seagoing vessel'.

The gradual disconnection of land from sea has spurred our fetishising of the latter, and a symbolic turning point seemed to come in 1924 with the opening of a fine new emporium for Liberty's on Regent Street, London. The business was founded in 1875 by Sir Arthur Liberty, a dealer in luxurious and rare textiles who had the misfortune to die a few years before the launching of his flagship.

Liberty's new building was dressed in the Arts and Crafts style then in vogue. The long elevation to Great Marlborough Street was a breath taking parade of timber framing in the manner of Elizabethan England and was intended to recall the stellar oceangoing of that age. For Arthur Liberty was a man who had conducted long-distance trade in silks and other fine fabrics for his entire career and was fascinated by ships and the sea.

He envisaged his flagship store along literal lines as just that: a ship fresh from the Orient docked audaciously on Regent Street, spilling the world's wares into central London. To deepen the authenticity of this illusion, he bought the timber carcases of two old sailing ships of the line then going spare: HMS *Hindustan* and HMS *Impregnable*. These ships he made shop-shaped; their timbers he had trimmed, spruced and erected as the frame of Liberty's. From the outset, this was a selling point, gleefully publicised from the moment the shop opened.

The idea was not only to graft ships into the shop, but also to craft a neo-Elizabethan architecture. 'There is a glamour about the lavish and stirring days of Henry VIII and Queen Elizabeth',[3] wrote Ivor Stewart-Liberty, the founder's nephew. In a eulogistic pamphlet about the building, he wrote of how the Tudor-Elizabethan architectural style is 'essentially English', bringing to mind 'a picture of those bygone days when the ancient guilds of the craftsmen and the merchant adventurers displayed, in the beautiful gabled buildings of old London . . . the treasures for which they sailed so far and endured so much'. The carving of the timbers was designed, according to Stewart-Liberty, so that it 'would not look out of place or give a discordant note, if it were put into a genuine Elizabethan house'.

In this, Liberty's stood apart from the Ramscraigs croft, St Nicholas's church and Stephen's cottage. For they were built or extended towards the dusk of sail, when sailing ships were still commonplace and the sea integral to the national timbre. But Liberty's opened during the steam age, when sail was just about furled and seafaring defined us less. That same year, the first Shipping Forecast was broadcast and the government procured *HMS Hermes*, Britain's first aircraft carrier. The former would become a much-cherished national staple, a sonorous fantasy of weather and hardship; the latter embodied the navy's ebbing prestige as airborne and submarine warfare increased in importance during the equivocal surface engagements of the First World War. In these, gigantic, evenly matched dreadnoughts exchanged long-range fire over the horizon, a spectacle far removed from the triumphs on which the navy's sense of itself depended. In the Battle of Jutland, the major sea-battle of the war, Britain's victory was equivocal, not complete.

Increasingly, seafaring was remembered, rather than practised. And as a result, a deep nostalgia for the age of sail set in. Buildings of ships' timbers became remarkable where once they were not. Liberty's is emblematic of this trend and, like a folk memory, ships' timbers have been variously claimed for buildings ever since. Now that we praise the noteworthy over the seaworthy, we treasure instances of the ship in the house.

14. Timbers of *San Josef* serving
as chancel ribs in the church of
St Nicholas, west Looe

Mast

Mast from the SS Great Eastern

MILLWALL, LONDON; LIVERPOOL
VICTORIAN

Incandescent after a spell in the coals, the rivet is chucked
by the boy to the riveters nearby. One of them catches it in a
bucket, then with tongs positions it over the hole in the overlapping hull-plates. Quickly, while the rivet is molten and malleable, the other two drive it in deeply, bashing the end into
a cap so that, when cooled, the rivet solidifies and holds the
plates fast together. There are many riveting gangs moving
along the scaffolds that teeter up the sides of this great ship.
Around 3 million rivets are needed for the vessel's outer skin.
From a distance, tossed to and fro, they glow like fireflies
against the black hull.

Despite the heavy industry there is a freshness to the
air. In the East London of the late 1850s, the land was half-
pastoral, half industrialised. Where this ship is being built, in
Millwall, on the south-west bank of the Isle of Dogs, the furnaces, engines and giant, half-formed structures make an infernal
contrast with the bordering pasture that has remained largely
unchanged since the time the Romans first arrived, over one
and a half thousand years previously; in just a few decades, it
will be swallowed up by manufacturing and housing. For now,
the hybrid terrain of the Isle of Dogs, hanging into the river's
bend like a tonsil, speaks of a nation in transition.

There is a sense, too, of a baton being passed between
the age of sail and that of steam. Unfinished, this immense ship
is as yet also unnamed. In size, there is nothing to equal her
at sea; scarcely anything on land is as big, except perhaps the
train-sheds of the new London railway termini. Accordingly,
she will first be christened after the oldest sea monster of them
all. Launched as *Leviathan*, this name - inspired by the biblical
sea monster - perhaps had uncomfortable associations with

the depths of the sea, not its surface, so she will soon be re-named *Great Eastern*.

To her creator, immensity was all: she was to steam to Australia and back without refuelling, undercutting and out-running the competition, and for this she would need hold-space as cavernous as an exhausted mine. She would be the famed engineer Isambard Kingdom Brunel's last and largest ship, and he took no chances with her colossal dimensions and the propulsion that she would require. He designed her as a hybrid, incorporating a propeller, paddle-wheels and six masts spaced evenly along the deck between her five smokestacks.

Of all the parts of a sailing ship, the mast seems most to-temic. Literally so, for, naked, it resembles an uncarved totem pole; symbolically, masts speak of a ship's origins, most resem-bling the tree-trunks from which came the rest of its timber; functionally, they make a sailing ship what it is, bearing the sails that give it motive power; and hierarchically, the main-mast was the point of orientation around which the ship was arranged. Crew roved before it, while the officers claimed the space 'after' or to the rear. But *Great Eastern* is a steamship, and her masts have a more ambiguous meaning, there to pro-vide auxiliary power but mounted like bygone relics between her paddle-wheels.

Shape-wise, at least, a mast is perhaps the largest of batons. Fittingly enough, *Great Eastern* was built on ground which had originally belonged to eighteenth-century mast-makers Ferguson and Todd. Their operation was conducted in timber sheds fronting the Thames, dominated by the sizable mast house and the ponds in which floated the unseasoned trunks awaiting their shaping. To the south lay spare ground, which in 1836 they sold to the Scottish engineers William Fair-bairn and David Napier who promptly established yards for building iron steamships. In them, twenty years later, con-struction began on *Great Eastern*.

On my desk, I have an old ship's rivet, a nub of iron mis-shapen from the moment it was hammered home. It looks like a black champagne cork. Rust has overcome it in places, but the marks of its molten manipulation are easy to see. That the rivets must be white-hot in order to join the hull's iron plates seems to symbolise the gulf between timber- and iron-shipbuilding. Fire is anathema to the former, fundamental to the latter. *Great Eastern* was born not in a forest, but in a furnace. Her dimensions exceeded those of any tree. Everything about her spoke of modern technology. And yet she straddled the ages of sail and steam, of masts and smokestacks. She illustrates a baton passing, but not yet passed. She would bring about a sea change in the standing of Britain's coastal places.

Iron has been prized since the earliest times, most obviously lending its name to the phase of prehistory it dominated. In Dr Johnson's definition of 1755 it is: 'a metal common to all parts of the world, plentiful in most, and of a small price, though superior in real value to the dearest'. And from the Iron Age onwards, iron has been present in ships, forged into the nails that fasten hull-planks, the pintles to hang the rudder, the anchors to fasten the ship to the seabed.

Yet, for a long time, this was the limit of its use at sea. Sailors were accustomed to iron sinking in the form of an anchor rather than floating in the form of a hull. And it was known to react with seawater. Johnson went on to describe it as 'not only soluble in all the stronger acids, but even in common water'.* In fact, 'iron sickness' was a real problem: if not adequately lead-capped, the iron nails would waste away under the corrosive power of the seawater, potentially springing thousands of tiny leaks in the hull as the sea jetted through the nail voids. And in a different, more strategic sense, iron was a shipbuilder's menace: its manufacture relied on wood in the form of charcoal and, as demand for iron rose during the eighteenth century, so ironmasters denuded England's forests of naval timber.

In this context, a wholly iron ship must have seemed

---

* By the sixth edition of the *Dictionary*, published thirty years later, this curious statement had been omitted.

unthinkable. Yet, in 1787, an iron boat was launched into an idyllic bend of the River Severn. Unfortunately, very little is known about the mysterious *Trial*, designed by John Wilkinson, an ironmaster of no little eccentricity who stood out amongst his peers for his delight in experimentation, manufacturing from iron a wide range of items that had never previously taken a metallic guise (his bed, his desk and, inevitably, his coffin). Wilkinson was ideally placed to attempt the construction of an iron vessel, which he conceived to ship materials between his Black Country furnaces.

Most likely *Trial* was modelled on the existing forms of river barges. She was around 70 feet long, six feet wide and could carry upwards of 32 tons of cargo; coincidentally or not, 32-pound cannonades reverberated through the trees as she was heaved into the river, disturbing a nearby rookery. This launch must have been a singular occasion, attended as it was by sceptical countryfolk and curious foundrymen, and taking place between two thickly forested riverbanks high up in the Shropshire tracts of the Severn. In my mind's eye she bore, with her triangular bows and simple, rectilinear lines, an uncommon resemblance to the Dover Boat, with which she seems to share a form of kinship (although *Trial* plied only inland waterways, never the sea).

While her form might have been more or less familiar to those onlookers on the riverbank, her buoyancy was not. Most of them thought – and were gleefully expecting – that she would immediately sink. We must therefore imagine a collective intake of breath as Wilkinson's men lugged and shoved her down the makeshift slipway, mightily expelled when she splashed into the shallows and bobbed there happily. What did the onlookers make of this? In the words of one contemporary, reporting *Trial*'s later arrival in Birmingham: 'it may seem somewhat strange that iron should swim . . .'[1]

As well as demonstrating iron's buoyancy, Wilkinson helped to resolve the contest for timber between shipbuilders and ironmasters. Further developing a process devised by Abraham Darby in 1709, who had successfully smelted iron

with coke instead of charcoal, Wilkinson managed to prove ordinary coal a viable fuel for iron manufacture. Abruptly, ironmasters left the forests for the navy's ends*; coal would prove,
in the eyes of Wilkinson and his fellow Shropshire ironmasters,
to be a virtually inexhaustible boon to their industry. Advances
such as these spurred the exponential growth of the iron industry, which in turn drove England's wider Industrial Revolution. And the ironmasters grew into industrial colossi. At the
peak of his operations, Wilkinson alone was responsible for an
eighth of England's total iron output, making nearly 16,000
tons per year.

From this point onwards, iron was cheap and capable
of being produced in large plates and, by the early nineteenth
century, successor craft to the *Trial* began to emerge on waterways throughout Britain. In 1810, an ironmaster named John
Onions built and launched *Victory*, a fifty-ton lighter used for
cargo on the Severn. Onions, in fact, built several iron barges,
one of which was prefabricated and shipped to a Mr Bishop,
who made it the first iron vessel upon the Thames. The advantages of their strong iron hulls were becoming more and more
obvious: once built, if competently navigated, they could go for
years without needing repairs.

Yet these early iron barges were, like *Trial*, anonymous
cargo-luggers, unlovely on the eye. Unsurprisingly, Onions'
Victory captured fewer hearts than her illustrious namesake.
Then, in May 1819, *Vulcan* was launched onto the waters of
the Clyde. Built as a passenger boat for the Forth and Clyde
Canal, *Vulcan* was iron-hulled in order to resist the ice which
frequently blocked the shallow canal and damaged the wooden
hulls of its traffic. But the importance of *Vulcan* lay in her
elegance: she was designed with a true ship's hull, as beautiful, in her way, as anything in the Nelsonic navy. Tellingly, her
lines were laid down not by an ironmaster, but by Admiral John
Schank, a Nelsonic officer who had designed improvements to

---

* Wilkinson helped the navy in other ways, notably by developing a process for casting and boring cannon without a single impurity.

navy ships. *Vulcan* would be almost the last work of this portly, seventy-something Admiral who, unlike so many of his fellow officers, was able to see a place for iron in a hull beyond that of a fixing nail.

Still, however, doubts about iron's seaworthiness persisted. Over thirty years would elapse between the surreal launch of the *Trial* and the first time an iron bow would head out to sea, in 1822. Equally surreally, this first seagoing iron ship emerged from furnaces far inland. She was built by another Shropshire engineer, Aaron Manby, who named her modestly after himself. Built at Manby's works in Tipton, South Staffordshire, the *Aaron Manby* was built, then deconstructed and shipped in pieces by canal to Rotherhithe, London, where she was rebuilt and launched in May 1822. She was captained by Manby's co-designer and business partner, Sir Charles Napier, a legendary naval officer of the age of sail.

After trials on the Thames, *Aaron Manby* departed London for Paris, taking around fifty-five hours to steam to Boulogne, and from there made her way up the river Seine to the French capital. Other Channel ships must have started at the sight of her (or him, perhaps, so thoroughly different was this vessel in motion and appearance to anything that had sailed before). Sitting low in the water, her flat-bottomed hull was 120 feet in length and unsettlingly featureless, sporting only a stark bowsprit, paddle wheels and a single, 47-feet-high smokestack. And unlike those other sailing vessels, which drew the wind's patterns upon the sea as they tacked towards port, *Aaron Manby* steamed single-mindedly to her destination.

Of course, other steamships had existed before her. Steam power had been tinkered with for centuries, but only in 1712 did Thomas Newcomen, a Dartmouth ironmonger and Baptist preacher, devise a way to make it continuously work an engine. Despite this momentous discovery, Newcomen's engine was used only to pump water out of mineshafts, without immediate application elsewhere. In 1755, Dr Johnson could still innocently define steam as 'the smoke or vapour of anything moist or hot', with no hint whatsoever of the way, a few decades later, the Scottish engineer James Watt would radically improve the efficiency of Newcomen's engine and so release the tremendous motive power of the Industrial Revolution. And it was only in the nineteenth century that steam power was loosed upon the sea, at first with equivocal results. The first successful steamboat, the *Charlotte Dundas* of 1802, looked like the awkward offspring of a mine-engine and a fishing smack.

But *Aaron Manby* was the world's first iron-hulled, sea-going steamship. She was driven by an oscillating engine, which featured a gasket rocking repetitively as if in stylised imitation of a boat's motion. Under its power she crossed the Channel a few more times, plying between Paris and London, demonstrating to sceptics the seaworthiness of metal and the wind-defying advantages of steam. She must have unshipped the senses of any sailors who boarded her, who would have seen plates where they expected planks, smoke where they expected canvas, and a lone funnel where they expected a mast.

Ore must be burned to make iron, to make hull-plates; once made, iron ships do not burn, unless sunk back in the furnaces from which they came. In this, the new iron ships were more impervious to the fires which threatened to sink their timber counterparts, immune to the terror that fire-ships evoked when set ablaze and shoved off towards timber fleets. Indeed, iron steamships *were* fire-ships, always carrying fire in their holds and smoking on the horizon.

'Next day, the 19th December, they burned the masts and spars . . . Next day the upper works disappeared, and the *Henrietta* was then only a hulk . . .' In Jules Verne's 1872 novel *Around the World in Eighty Days*, Phileas Fogg and his companions steam frantically across the Atlantic on their last sea-crossing, bribing the *Henrietta's* crew to alter her destination from Bordeaux to Liverpool. Such is the pace at which they steam that they run out of coal half way across the Atlantic; in response, Fogg buys the ship from the captain and then, to keep up their speed, burns everything of her except her iron hull.

Here, satirised, is a problem which assailed the early steamships: the quantity of coal required to steam any distance would consume all the space available for profitable cargo or passengers, making this form of power less economically workable than sail. *Aaron Manby* could carry enough to cross the Channel, but the prospect of a profitable Atlantic crossing (or further) under steam looked remote. Yet little more than a decade after *Aaron Manby*'s maiden voyage, the commercial viability of steamships would be proven by a young engineer with no experience whatsoever of naval architecture.

Isambard Kingdom Brunel was aptly named. For the scale on which he worked was that of a kingdom and his rule of schemes was absolute (and the middle name is key: 'Isambard Brunel' has not quite the same ring). Roving easily across disciplines, designing bridges, railways, ships, buildings, he frequently created superlatives of their kind. Few among his Victorian contemporaries could equal his vision or, perhaps most importantly, his flair in projects of the hugest proportions.

In 1835, Brunel, then just twenty-nine, was engineer of the Great Western Railway under construction from London to Bristol. In a board meeting with the directors, he suggested the line be extended to New York by a steamship. Steam, rather than sail, would provide quicker, more reliable motive power and would dovetail in spirit with the railway's steam locomotives. But no-one thought a viable transatlantic steamer possible.

In defying the sceptics Brunel, typically, theorised that size was all. Building on the work of earlier physicists, he came to understand that when a ship is scaled up, its fuel requirements increase as a square of its dimensions, while its hold space increases as a cube. In other words, larger ships are more fuel efficient, giving them proportionately more hold-space for paying passengers. His proof was SS *Great Western*, a wooden-hulled paddle steamer built in Bristol and launched in 1838. In one respect, she was not as futuristic as *Aaron Manby*, for she carried a supporting cast of masts and sails. Yet on her maiden voyage she crossed the Atlantic with passengers and fuel to spare, proving that steam could pay.

Of all his ships, Brunel is perhaps best remembered for SS *Great Britain*. Built in 1843 as a sister to *Great Western*, she was the first large iron vessel, 322 feet long to *Aaron Manby's* 120 feet. Formed of wrought-iron plates riveted together to an unprecedented length, her elongated hull, like *Vulcan*'s, was reminiscent of the traditional sailing ship, yet iron-made: old form, new fabric. She, too, carried masts, but these were demountable and were far overshadowed by her novel screw propulsion. After a long, successful transoceanic career, carrying passengers to both the Americas and the Antipodes, she was hulked and then scuttled in the Falkland Islands, where she was seen during the Second World War by the lighthouse keeper W. J. Lewis: 'at rest at last, very old and well worn, with water in her hull but still lovely in the peace of Sparrow Cove'.[2] Now all that remains of her there is a lone mizzenmast; her hull floats in Bristol, splendidly restored.

In November 1857, Robert Howlett took the famous photograph of Brunel posing before a drum of stupendous iron chains, cigar in mouth, hands jammed into pockets. By this time, at fifty-one, he had laid the main lines of the Great Western Railway, which incorporated the ingenious Maidenhead, Chepstow and Saltash bridges (among other architectural marvels) and sent forth two pioneering steamships into the world. It's this photograph by which he is often recalled, a prodigy in muddy trousers, dwarfed by the scale of his creations.

And the image not only immortalises Brunel, but also the ship on which he was then working. For he stands in the Millwall shipyard where SS *Great Eastern* was under construction. This, his next and final ship, he conceived on a far grander scale than her sisters. Designed to be fuel-sufficient for the Antipodean run, she would be by far the largest ship the world had ever seen: twice the length of SS *Great Britain*, at 692 feet. A measure of her scale is given by the size of the chains in the background of the photograph, forged to lower her stern into the Thames. At the time of the photograph, her launch was imminent. There had already been difficulties with so colossal a build; characteristically of Brunel, her scale and displacement were untested.

Like *Trial*, then, launched seventy years previously, *Great Eastern* was an unknown quantity, although the quantity of iron in her far exceeded her predecessor. A sense of uneasy anticipation attended both launches, a sense of being about to see history made, or a fool made of a shipwright. Both were wonders of their day. Both Wilkinson and Brunel were imperiously confident in their calculations, but yet to see empirical proof of them. In that photograph of Brunel, it's hard to read the expression on his face. I daresay it's that of a shipbuilder yet to see his vessel float.

Pickaxed from the ironfields, the grey rocks are lifted from the mines and then hauled by train to the works. In a blast furnace, the iron liquidises from the rocks and outflows into a mould of rough ingots connected to a central runner like suckling piglets. It's left to cool in this intermediate, 'pig-iron' state – still too slag-ridden for the finished product – then thrown into the puddling furnace, where it reddens again and is stirred with poles until the impurities are burned off and the iron globules can be fished out. These are combined and thrust into yet another furnace, from which they emerge a white-hot mass to be rolled into hull-plates.

Fittingly, 'Isambard' derives from a Germanic word for iron; *Great Eastern* would be the largest iron building then raised. And the circumstances of her birth, conducted on so vast a scale, emphasised the differences between ships industrialised and ships carpentered. Iron did not grow upon the surface of the nation but was delved out of its guts. Worked not by hand but by heat, it passed through cycles of amazing incandescence to arrive at a state ready for shipbuilding. And, in *Great Eastern*, the sheer quantity of it was overwhelming: 30,000 iron hull-plates passed between the rollers.

(Put like this, iron appears a strikingly unsustainable material. Not only were vast quantities of coal burned to attain the temperatures at which it could be formed but, like the coal, the stocks of raw ore were finite. At the height of the Industri- al Revolution, mine-riddled Britain must have seemed like a nation devouring itself. And forests can regrow, whereas mines remain spent.)

However, it was likely a softer, more seductive metal that had seeded the ship. In early 1851, gold was found in Aus- tralia, news of which electrified the world. An international rush set in. Suddenly, Australia ceased to be thought of only as the world's largest and remotest penal colony, and instead came to be seen as a beckoning land of promise. Across the world, aspiring panhandlers raised what capital they could to cross the oceans and arrive in their thousands at the port of Melbourne. From there, they trekked inland to vast, tempo- rary claim-cities. But there was an ironic quality to this gold. So many sought it out that proportionately few were enriched. The real gold, it might be said, was the lucrative carrying and maintenance of this stream of proto-prospectors.

Nevertheless, this Australian gold rush turned heads to- wards the east. Shipping companies sought to capitalise on the emigration and the wider trade with Asia and the Antipodes. In 1852, Brunel designed two steamships for the Australian Royal Mail Steam Ship Company. *Adelaide* and *Victoria* were reitera- tions of the successful formula he devised for *Great Britain*, but their careers are somewhat obscure; the following year their parent company lost its lucrative mail contract and folded. Meanwhile, it had already occurred to Brunel that a ship even larger than *Great Britain* would be better suited to the Antipo- dean passage. Only a boat of huge size could carry enough coal to steam beyond the range of rival steamships. If successful, it would far outrun its competitors under sail, for whom the journey averaged one hundred days. Under constant steam, it could take just thirty.

It happened that a new shipping company with Antip- odean ambitions, the Eastern Steam Navigation Company (ESNC), had been formed in the early 1850s. However, it failed to secure the requisite Royal Mail contract – essential for subsidising the high operating costs of the voyages – which went instead to the rival Peninsular & Oriental Steam Navigation Company. The ESNC was, therefore, a shipping company without ships, facing dissolution until Brunel sent a paper pitching his leviathan; the company was deeply intrigued. Certainly, the outlay to build such a vessel would be enormous, but it might be recouped on a voyage carrying thousands of paying passengers without the costs of constant recoaling.

But while only Brunel was confident enough to build with iron on this scale, he realised that he could not build *Great Eastern* solo. In any case, all his previous vessels had been collaborative affairs: he supplied the outlines, the general principles and the captaincy of the projects, while other shipbuilders and engineers helped him to execute them. And there was only one place in the world where a ship of this size might be built.

To realise *Great Eastern*, Brunel entered into partnership with John Scott Russell, a Scottish physicist, naval architect and shipbuilder who pioneered refinements to the shapes of ships' hulls to minimise water resistance. After an eclectic start to his career working, successively, as Professor of Natural Philosophy, experimental barge-builder, manager of a Greenock engine works, editor of the *Railway Chronicle* and secretary of the Royal Society of Arts,* Russell bought William Fairbairn's shipyard at Millwall in 1847. There he built, among many others, Brunel's steamships *Adelaide* and *Victoria*, and developed a mastery of iron fabrication.

It was Russell who presented the concept of *Great Eastern* to the board of the ESNC (Brunel was unable to attend); it was Russell who then tendered successfully for the job following the board's acceptance of the idea. And it was Russell who subsequently designed the ship's hull and its paddlewheel engines, making *Great Eastern* as much his ship as it was Brunel's. On 22 December 1853, Russell signed a contract providing for the 'construction, trial, launch and delivery of an iron ship of

---

* In which capacity he was instrumental in the concept of the 1851 Great Exhibition.

the general dimensions of 680 feet between perpendiculars, 83 feet beam and 58 feet deep according to the drawings annexed, signed by the engineer I. K. Brunel'.[3] So vast a ship would not fit into Russell's yard, so he expanded into the empty, adjacent shipyard formerly owned by David Napier. And as the ship's length was almost equal to the Thames's breadth at this point, if she were launched stern-first into the river (as was customary) she was likely to plough into the Deptford shore. So, she would be built and launched sidelong.

On May Day 1854, Russell's men laid down her keel-plate. From this iron spine grew two iron hulls, one inside the other, of tapering thickness: an inch at the base, three-quarters of an inch below the waterline and half an inch at her upperworks. Inside, bulkheads sectioned the hull into water-tight compartments. These innovations were designed to contain the invasion of the sea should the ship suffer any gashes through collision; to stave off corrosion, the seaward ironwork was everywhere painted black.

Over four exhausting years, *Great Eastern* grew inter-mittently from its keel-plate. She was then unparalleled: long as a highway, broad as a warehouse, tall as a cathedral, bigger than any contemporary dock. She towered over foregrounded houses when seen from Millwall marshes, while from the Dept-ford side she looked like a riveted cliff-face. The spectacle was captured in the photographs of Robert Howlett, commissioned by Brunel to document the build and publicise the ship. Some-how, Howlett managed to sink the vast hull into a camera lens. In his shipyard scenes, it sits massively amidst an incredible chaos of timber, bricks, chains, ropes, mud and foreshore ooze. It looked like what it was: a build of ad hoc uncertainty, follow-ing no precedent.

Howlett's photographs show how, despite her size, the ship's hull-lines were quite beautiful. Her tonnage required all Russell's ingenuity in overcoming water-resistance. When viewed broadside-on, the impression was simply of gratuitous length (necessary to hold all the coal); however, when viewed obliquely, her bows tapered slenderly to a spear-point and her

15. *Great Eastern*, elevation from mid-
Thames, Robert Howlett, 1857

stern swept elegantly to a curve. Between them, five huge boilers, coalbunkers and their funnels occupied the hull. These
generated the steam which drove the paddle engines, situated
just forward of the centre, and the engines which spun the two-
feet-thick propeller shaft. Beyond these operational areas, the
hull held a series of hall-like voids destined for cargo, accommodation and passenger saloons.

At 11,000 tonnes, the hull's great weight pressed both
on the foreshore of the river and the foreheads of its builders. Brunel and Russell fell out. Their agreement was just vague
enough for them to interpret it differently. While Brunel demanded total control down to the last rivet, Russell preferred a
freer hand in the detail. Increasing discord between them cul-
minated in strongly worded correspondence from which there
was no return (at one point, Brunel threatened that he would
like to give Russell a flogging). Alongside, the shipyard was assailed by a steady stream of petty thefts and then a damaging
fire, which burned much of Russell's facilities but thankfully
spared the ship. And then in 1856 Russell, bedevilled by creditors, declared himself bankrupt.

There were many reasons for this crisis: illiquid capital
tied up in shares, lax management of the build and its logistics,
and over-commitment to building other ships at the same time
(Russell's acumen, as his eclectic career hints, was more in-
tellectual than administrative). And there were wider factors,
too: the Crimean War, for instance, had disrupted the global

economy in such a way as to raise the wages of the labour force. Brunel and the Company rushed to take possession of the ship before Russell's creditors did; only after a bout of wrangling with Russell's bank could they lease back the yard in order to finish the job.

By 3 November 1857, they were ready to launch her (hull and engines complete, interiors and exterior upperworks outstanding). Brunel would have liked more time but had reluctantly agreed to this date because it cost an eye-watering £1,000 per month to lease the shipyard from the bank. He was not sure what would happen. Despite the ship's colossal size, he hoped for a discreet launch, conducted with minimal publicity and with silence in the shipyard so that orders could clearly be heard. However, unknown to Brunel, the stricken ESNC had sold 3,000 tickets to the event. Spectators packed the shipyard and thronged the opposite shore. Charles Dickens was there:

> A general spirit of reckless daring seems to animate the majority of the visitors. They delight in insecure platforms; they crowd on small, frail, house-tops; they come up in little cockle-boats, almost under the bows of the great ship.[4]

No-one had ever attempted to move 11,000 tonnes at once. Brunel's solution was to push the ship with two hydraulic rams and pull her with barge-mounted steam winches; under these forces, and checked by two great windlasses, she was supposed to make a stately descent upon two massive cradles down into the river. It was among the greatest technical challenges of a career not lacking in them, and overshadowed any other matters. When presented with a possible list of names for the ship, Brunel is alleged to have remarked: 'call her Tom Thumb if you like' (his pet name for her, the Great Babe, seemed unsuitable).

Just after midday, Henrietta, daughter of the ESNC chairman, named the ship *Leviathan*; just after her christening, the launch commenced. At the windlasses, men tensed; the river lapped at the slipway foot. All was silent in the yard, then the chains hauled and the rams heaved. *Leviathan* moved a few feet – then halted.

Some way out of the city of Liverpool, I walk through drizzle along streets unfamiliar to me. Chill mist spills from the river Mersey over a low suburb, an acreage of red brickwork and battered shopfronts with Anfield at the epicentre. Alongside scarved Liverpool supporters I make my way towards the stadium, which dwarfs the two-storey houses lapping up against the sides. You might think of it as navel architecture, this blocky form centring a circle. Reaching the stadium's southern corner, I find the mast I seek. Here is all that remains of *Great Eastern*.

Great Eastern was given six masts, named for the days of the week. Monday, Tuesday, Wednesday, Thursday and Friday were made of iron, while timber formed the Saturday mast. To Saturday the ship's compass was fixed; iron would have seduced the needle and therefore been an improper material. And, neatly, I stand before it on the day for which it is named. Now cast into a concrete floor, stayed with steel cables, the Saturday mast spikes upwards but falls short of the stadium's height. Stepping back to view the two together, it looks like a mooring-post to which the vast bulk of the stadium might be tethered.

At first it seems odd to see *Great Eastern*'s last mast juxtaposed with such a massive structure. A mast should be the tallest element of a view, standing proud of its surroundings. Here, it's sunk below the apex of Anfield's roofline, in a strange, hierarchical inversion; here, the mast has ceased to be a yardstick for height, and instead communicates the stadium's depth. And yet there is something apposite here, for depth charges the story of Brunel's biggest ship.

Great Eastern failed as a liner but succeeded as a line-layer. Between 1865 and 1874, she would lay a series of submarine telegraph cables between the world's continents. As the largest ship afloat, she could carry thousands of miles' worth of cable and coal over distances that would tax even a phalanx of smaller cable-ships. In doing so, she would help to wire together the world, making possible instant international communication. But in paying out these cables into the depths of the ocean, she would pay off ships as carriers of news. For the telegraph lines would pass between capitals, bringing it straight into the centre, bypassing the vessels in which it had previously travelled and the harbours at which it would first have been heard.

Suddenly alone outside the ground, I again run my eye up the Saturday mast, displaced from the ship into this most unlikely context. A sort of displacement of ships, indeed, was what *Great Eastern* helped to bring about; or, at least, a kind of demotion of them in the national psyche. For information – intelligence – is a most prestigious cargo. Without it, ships were conveyances only, the sights of their masts on the horizon promising only goods or people, rather than the news of the moment.

Yet masts would continue to afforest harbours, estuaries, coastlines. Sail had enjoyed a late triumph in the mid-nineteenth century with clipper ships, extraordinarily fast vessels designed to run perishable cargoes from the Far East back to lucrative Western markets; *Cutty Sark* is the only survivor of these thoroughbreds (the rivet on my desk comes from her; she was wind-driven but her hull was a composite of iron and timber). Eventually, however, steam surpassed them. Sailing vessels held a waning role transporting low-value bulk cargoes, but that was all.

From the belly of Anfield comes a deep roar. I take one last look at the Saturday mast, then walk back into the mouth of the city. *Great Eastern* was, emphatically, a ship of iron and steam. That her only remnant should be a mast pinpoints her significance. For Brunel, with characteristic foresight, had dreamt up and executed *Great Eastern* long before the heyday of steam; sail still had a century of vitality ahead before Britain's seas were dismasted. This mast is a token of how *Great Eastern* spanned the two; but more particularly, it stands for how, in her cable-laying work, she forever changed what a mast could promise.

Launch day had been a failure: in the end, it took three months to shove the ship far enough down the slipways to the point where the Thames, in spate, could finally float her off. That was in January 1858, and she was then still only a hull with engines. Fitting her out as a liner took over a year. Her first foray into the sea, in August 1859, was marred by a huge explosion off the coast of Hastings which felled her forward smokestack and destroyed the Grand Saloon. The cause was a single valve that was closed when it should have been opened. Hearing of this, a sickening Brunel died. As if this wasn't bad enough, her first commander, Captain Harrison, drowned five months later in the harbour at Southampton.

In the eyes of the public, she was an unlucky ship. At least, that has to be one explanation for the fact that on her first Atlantic crossing she carried just thirty-five paying passengers (when she had been designed for thousands). As it turned out, she handled reasonably well despite her size: to one passenger on board, she moved in a storm 'like some accomplished swimmer, who sweeps forward gracefully hand over hand'.[5] Arriving in New York for the first time in July 1860, she caused a sensation. Thousands of people paid fifty cents each to tour the ship. Indeed, as future years would show, this was seemingly the only way to turn a profit from her. People saw her as a spectacle, rather than a reliable means of transport. In ports across the world she transfixed the crowds, but comparatively few actually travelled on her. She really belonged to the golden age of the ocean liner, but it had not yet arrived. In the early years of the 1860s she made several more Atlantic crossings, with a more profitable passenger list, but accumulated storm damage and gouged open her outer hull on a reef outside Long Island. By 1863, facing increasingly fierce competition from other transatlantic ship operators, her owners found themselves sunk in debt. She was laid up,* gathering barnacles,

---

* I can find no records of where she had previously been repaired – possibly it was Milford Haven, where she was later mothballed for a time.

while they cast about for her future. However, there was an unforeseen job to which she was fortuitously well suited.

Submarine telegraphy was an enterprise as extraordinary as the ship herself. A single cable, barely a handspan's diameter, would be lowered down thousands of fathoms to snake across the ocean floor. Through the cable ran a core of the purest copper thousands of nautical miles long, insulated with gutta-percha (gum from Sumatran forests) and shielded by wire armouring. It would pulse messages from continent to continent, far outrunning the wakes of the ships far above. In the later words of poet Rudyard Kipling: 'here on the tie-ribs of earth / Words, and the words of men, flicker and flutter and beat . . .'[6]

In 1850, a primitive cable was slung between England and France across the floor of the English Channel. In 1858, a hardier, more intricate cable was successfully laid between Ireland and Newfoundland, this being the shortest distance between the (then) United Kingdom and the United States. Soundings taken by an American naval officer, Matthew Maury, showed the ocean floor along this route to be so flat and comparatively shallow that it seemed, like *Great Eastern*, providentially well suited for a submarine cable; Maury nicknamed it 'Telegraph plateau'. An Anglo-American consortium, headed by telegraph pioneer Cyrus Field, established their cable along this route, laid by two converted warships, HMS *Agamemnon* and USS *Niagara*. In July 1858, they met in the mid-Atlantic to splice their cables together and then paid out the whole as they steamed home.

A month later, Queen Victoria and President James Buchanan conversed from their countries in the space of a day. Possibilities – in government, diplomacy, commerce, science, in all spheres, in fact – rapidly unfurled. The enterprise revealed how to the Victorian mind the oceans – increasingly sounded, quantified, intervened in – were no longer the sublime abysses they had always been. And then, a few weeks later, the cable failed.

Possibly this was due to an over-zealous electrician sending a powerful bolt of current down the line to chivvy a flagging message. Messages traversed this wire slowly, at the rate not of seconds, or even minutes, but hours per word. Or an older section of cable, spliced into the new, may have harboured compromising weaknesses. Whatever the cause, the failure was so sudden and so complete that some were convinced that it had all been a hoax, a notion assisted by the fact that the cable had not had time to open to commercial traffic and had conveyed only the utterances of heads of state and telegraph clerks.

To Cyrus Field, though, this short-lived cable proved that it could be done. But investors were wary. It took years to raise enough capital and secure backing from the British and American governments for a second attempt. By 1865, Field was in business with John Pender, a Scottish textile magnate who had amalgamated several cable manufacturers to form the Telegraph Construction and Maintenance Company. He was also *Great Eastern's* new owner.

After her faltering efforts as a passenger liner had brought her owners nearly to bankruptcy, she had in 1864 been offered for sale in Liverpool. Perhaps her aura of misfortune crept into the auction room, for few serious bids were received. In the end, she was offered for sale without reserve and bought for around a quarter of the cost of her construction. Pender jointly owned her with two railway magnates, Daniel Gooch and Thomas Brassey, who, in the enterprising style of the age, had put up their own money to buy *Great Eastern* in order to lay the cable. In a way, they had little choice but to buy her. No purpose-built cable ships then existed, only awkwardly adapted vessels, none of which could have single-handedly carried the 2,500 nautical miles of cable (and the requisite coal) to reach between Ireland and Newfoundland.

Spring 1865 found *Great Eastern* moored at Sheerness, the limitless spool of Atlantic cable being painstakingly coiled into her three new cable-tanks from two old warships loaned by the Admiralty: HMS *Amethyst* and HMS *Iris*. They had seen much, these venerable craft, and they were among the last of their kind. The career of HMS *Iris* was particularly noteworthy: built in Plymouth in 1840, she had cruised off West Africa, suppressing the slave trade, and had fought engagements in the waters of New Zealand. Both ships, by now hulks, were built along lines reminiscent of the navy's Georgian pomp; the sight of them subserviently moored under *Great Eastern*'s iron lee poignantly captured the changing times.

After the cable was taken aboard, *Great Eastern* was loaded with coal and then steered to Berehaven, a deep-water port on a blade of the southern Irish coast. In the meantime, the shore end of the cable was laid out of Foilhummerum Bay, Valentia Island, to a buoy approximately 70 nautical miles offshore, where *Great Eastern* would pick up the slack. On 23 July, she spliced this cable with the one coiled in her tanks. And then, attended by another pair of warships, HMS *Terrible* and HMS *Sphinx*, she steamed westward.

Laying a submarine cable sounds easy: simply pay it out over the stern and watch it disappear into the ship's widening wake. Actually, it was trickier than this. For one thing, the speed of the lowered cable and that of the ship had to be satisfactorily aligned: too slow, and the cable risked gathering indolently on the ocean floor; too fast, and the cable might never settle on the bottom but instead be hauled through the depths. For another thing, the cable could not simply be left to slip over the stern of its own accord. It needed to be unspooled at a uniform rate on a uniform path between the tank and the stern. Ever erring between taut and slack, the cable had to be constantly watched, fussed over, reguided. On several occasions, the wire armouring the cable pierced the insulation (detected by a device attached to the cable called a mirror galvenometer), meaning the paid-out lengths had to be recovered and repaired. And then, in early August, the cable snapped and whipped, spaghetti-like, into the maw of the sea.

Although always a very real risk, this was the worst thing that could have happened. The costly cable now lay on the ocean floor, at an estimated depth of 2,000 fathoms away from the ship, a distance equivalent to over three kilometres above ground level. Desperate, the crew decided to attempt an unprecedented recovery. *Great Eastern* steamed a little way back along the cable's route and was positioned to drift and crisscross it. Early that evening, a grapnel attached to a wire rope was thrown overboard, taking over two hours to sink 2,500 fathoms down to the seabed. The ship drifted through the night until early the next morning, when, extraordinarily, the lost cable was hooked. Slowly, carefully, the crew began hauling it in, only to have the wire rope break; they had lost the cable again. For days they drifted and steamed in fog, their precise position lost, trailing ropes and grapnels along the ocean floor. Unbelievably, they again hooked the cable, but then lost it for a second time.

Recovery was next attempted the following year. A new company was formed, fresh finance found and the ship re-coaled and re-cabled. This time, it was a complete success. In a summer fortnight, *Great Eastern* steamed across the Atlantic, smoothly paying out her new cable onto the ocean floor. It ran from a cable hut in Foilhummerum Bay, Ireland, to Heart's Content, Newfoundland, and opened almost immediately to commercial traffic. After this, it seemed as though *Great Eastern* had finally found a role. In 1869, she was chartered by a French telegraph company to lay a series of cables between France and America. And, that same year, plans were afoot for her to lay a line to the east.

To connect Britain to India, John Pender had formed a series of cable companies for different sections of the route: it would join Bombay and London via Aden, Suez, Alexandria, Malta, Gibraltar, Carcavelos (Portugal) and Porthcurno (Cornwall). Apart from an overland section through Egypt, all these cables were submarine; the shorter sections were laid by a fleet of steamships converted for the purpose. The longest and most momentous stretch – symbolically, because it touched the Indian subcontinent – from Aden to the Indian mainland, would be laid by *Great Eastern*.

Portland, Dorset, is a quarry-gouged isle dangling into the English Channel. From here came the fine white stone for the most prestigious buildings in Britain (and it still supplies the stone for their repair); from the quayside here, in November 1869, came the last shovelfuls of coal *Great Eastern* required for her maiden voyage to Bombay. As the last shower of black rubble cascaded into her bunkers, observers may have thought they were seeing her phantom. For she was now painted a splendid off-white, having originally been launched as a black ship; her life up until this point had been chequered.

The reason was practical: in the heat of the Indian Ocean, an iron ship painted white would be around twenty degrees Fahrenheit cooler than one painted black. Not only would this be more comfortable for the crew, but it would also help to preserve the precious cable, susceptible as it was to heat in its rubber and copper essentials. And, in a way, her new livery was a symbol of her newfound success.

Inside her, in three circular tanks, lay coiled 2,735 nautical miles of cable. It was not all of uniform appearance: once laid, the cable's deep-sea parts lay relatively undisturbed, so required comparatively little protection and were relatively slender, while the shore-ends, which might be snagged by anchors or damaged by remora or crustacea, were thicker and more resolutely armoured. It lay there silently amidst the ship's other occupants: 35 bullocks, 120 sheep, 50 pigs and 1,200 chickens. Such was the fuel for her people; to power herself, she carried over 10,000 tonnes of coal.

After calling at St Vincent, the Cape Verde Islands and the Cape of Good Hope, she reached Bombay in January 1870. After coaling, she began cable-laying on Valentine's Day, steaming westwards across the Indian Ocean. The laying passed without a hitch.

By midsummer 1870, the line between Britain and India was complete. On 23 June, John Pender held an inaugurating soiree at his house in Arlington Street, Mayfair. Around 700 people mobbed the house, including the Prince of Wales and sundry other luminaries. Not bad for the son of a Dunbartonshire textile merchant. But such a stellar gathering could only have been drawn by a truly extraordinary event. Pender had converted part of his saloon into a telegraph-office and from there, at the beginning of the evening, a question was posed to India: 'How are you?' It must have been an unforgettable moment for those gathered in that opulent saloon, oppressed by the high summer heat, squashed amongst glittering uniforms, waiting for this seminal response. It took a few minutes to traverse the distance from the Himalayas to Piccadilly: 'All well'.[7]

Swaying gently in his cabin, the journalist wrote: 'There was a wonderful sense of power in the Great Ship and in her work; it was gratifying to human pride to feel that man was mastering space, and triumphing over the winds and waves; that from his hands down in the eternal night of waters there was trailing a slender channel through which the obedient lightning would flash for ever . . . binding together the very ends of the earth.'[8]

William Russell was aboard *Great Eastern* on her inaugural line-laying expedition in 1865 and was carried away by what he saw. But Russell had also been an earlier beneficiary of the possibilities opened up by submarine telegraphy. For he had been correspondent for *The Times* during the Crimean War, despatching reports from the front through overland telegraph lines and the cross-Channel cable laid in 1850. This cable enabled real-time war reporting, making the Crimean War the first to be fought in public. Russell's British audience eagerly followed the progress of the war through his reports.

Suddenly the world was more visible. Cables quickened the flow of news into Britain, galvanising public opinion and often leading to changes of policy or even of government. Outcry at Russell's reports of army mismanagement led to the collapse of the Aberdeen administration responsible, and the ascent of Lord Palmerston. A few decades later, during the siege of Khartoum, the mercurial General Charles George 'Chinese' Gordon was able, through copious telegraphy, to openly defy his superiors with public opinion on his side.

But the cables were not wholly disadvantageous to the government. They enabled a new-found control of armies and diplomats overseas, centralising power and privileging access to official information. Wars could be averted through quicker and more accurate knowledge of nations' intentions and military capacities. And, as news-gathering pioneer Paul Julius Reuter discovered, the cables were able to simultaneously swell and satiate demand for financial news, which in turn paved the way for the increasing globalisation of the world's economies.

International news now raced through copper across the seabed, between coastal telegraph offices, and was then relayed by clerks through overland cables to London – which, thanks to these new conduits, gained a new importance as the epicentre of information. Almost at a stroke, ports and harbours, once so important on the world stage, had been diminished.

In September 1580, Plymouth was the first to hear of Drake's circumnavigation of the world. Two centuries later, in 1771, Deal, Kent, welcomed back Captain Cook after he had become one of the first Europeans to reach the east coast of Australia. On Sunday, 25 November 1781, 'official intelligence of the surrender of the British forces at Yorktown arrived from Falmouth at Lord George Germain's house in Pall Mall . . .'[9] (Yorktown presaged the breakaway of Britain's American colonies and their eventual independence.) It was also at Falmouth, on 1 November 1805, that Lieutenant John Richards Lapenotiere landed HMS *Pickle*, having sailed directly from the Battle of Trafalgar with news of Nelson's victory and subsequent death (he reached London with the news five days later, after a non-stop race overland). Elsewhere, in midsummer 1815, news of Wellington's victory at Waterloo was first heard at Broadstairs, Kent.

Of course, not all coastal ports shared in the glory of such news. But the crucial thing is that, before the submarine cables were laid, they all harboured the potential. Unfavourable weather or a need for speed could force ships carrying it to run for the nearest port, however insignificant, on the closest accessible stretch of British coastline. And even if the despatch was not officially proclaimed at the moment of arrival, there were plenty of ways in which it could leak out through the disembarkation of the crew. From the smallest harbour to the largest port, places other than London, places we might now think of as lacking promise, lay open to this prestige. Now, they were bypassed.

Sail had trafficked news in a looser, more local way; news, intelligence and information were still prized, of course, but there existed no single global conduit through which access might be ever more rigorously controlled and privileged. After the cables were laid, masts no longer meant news. Of course, they still meant many things: safe passage of a cargo, or a loved one, and mail was still carried upon ships. Coastal places were not killed off instantly. Many continued to prosper well into the twentieth century, sustained by their other roles as pleasure-resorts, fishing ports, liner terminals. Only, everyone knew there had been a sea change in their standing.

Evening draws down over the redbrick streets leading to the docks. Lamplights come on all at once, pooling yellow light upon the pavements. I walk from Anfield down towards the Albert Dock. Contrary to what the Saturday mast might suggest, *Great Eastern* does not lie buried beneath Anfield stadium. And though the mast peculiarly captures one facet of her career, it is a most unsatisfying representation of the whole. What of the rest of her? What of the iron which so defined her and the steam engines by which she was propelled?

After laying the cable between Bombay and Aden, she spent some time reposing in the Mersey, before laying several more Atlantic cables between Ireland and Newfoundland. By 1874 she had finished laying these, her last lines; indeed, the previous year *CS Hooper*, the world's first purpose-built cable ship, had been launched from a Newcastle shipyard. *Great Eastern* increasingly appeared obsolete for this, or any other task. Of unique size and configuration from the first, as the years advanced these qualities ceased to be novel and came to be encumbrances. Other ships had better engines, were more manoeuvrable, were faster - though none, certainly, were as big.

From 1874 she would spend twelve years deep in retirement at Milford Haven, the westernmost Welsh port, whose town was laid out in the 1790s by a French shipbuilder (although the street pattern is sadly conventional). Not much happened to her here, beyond the seizing up of her engines, a few repairs and an odd incident in which she was trapped in a dock that had been built around her with too small an entrance – she was floated out only after her paddle-wheels had been removed. From the early 1880s she was put up for sale numerous times, though no credible buyer could be found. Eventually, she was bought by a consortium of traders for use as a coal-hulk in Gibraltar – but before she departed for this sad end, she was chartered for a year by Louis S. Cohen, the manager of the Liverpool department store Lewis's, for an even more ignominious errand.

She arrived in Liverpool on 1 May 1886, a tragic sight. Cohen had had her considerable sides emblazoned with adverts for Lewis's and anyone else willing to pay; he fitted her out inside like a fairground, with stalls, sideshows and areas set aside for trapeze-artists and juggling acrobats; it was noted, with distaste, that 'a circus will probably be established in one of her cable tanks'.[10] So outfitted, she attracted nearly half a million people. Abandoning the Gibraltar scheme, her owners took her to other ports to replicate this success but these attempts flopped. Eventually, in 1887, she was put up for auction one last time, and failed to attract a bidder. In the end, her now-desperate owners quietly agreed to sell her to Henry Bath and Son, shipbreakers, for a paltry £16,500.

Great Eastern might have been built in east London, but her true origins were northern, in Lincolnshire ironfields and Rotherham furnaces, from which came, respectively, her ore and then her hull-plates. It had been a long journey south, dragging those thousands of tonnes to the Thames, which, from the 1850s, was the focus of a brief but extraordinarily intense bout of iron shipbuilding, before an economic crash of 1866 wrecked the fortunes of its yards. By 1869, John Scott Russell's old shipyard was but a grassy waste.[11] As demand rose for large iron and, later, steel-hulled ships, the focus of shipbuilding would swing up to the north, to be closer to the raw material and the expertise required for its working.

The biggest and best ships of the late nineteenth century would issue from northern rivers: the Clyde, the Mersey, the Tyne. And those same ships, or their kinships, would at the end of their lives be dismantled on their banks. Among them was *Great Eastern*. In August 1888, Brunel's Great Babe made a stately glide up the Mersey, watched by thousands of people on both shores. She was to be beached on the Wirral side and, a little over an hour after high tide, she crunched to a stop on the foreshore.

And there she was broken up. Newspaper reports of the time praised the quality of her ironwork and the fastness of her construction. Those rivets, which had so easily pegged together the hull-plates when malleable, now formed a rock-solid wall. Unable to undo the hull by chiselling them out, the breakers resorted to a wrecking-ball designed to spring them from their sockets (this seems to be the first documented use of a wrecking-ball in Britain; *Great Eastern* was ever an engineer's inspiration). It had taken four years to build the ship; it took two hundred men two years to take her apart; it must have been strange for Liverpudlians, who had so often seen her entering the Mersey, to watch her slow-motion demolition. By July 1890, she was a vast void between skeletal traces of her bows and rudder-post. By September, nothing more could be seen.

Riverside, now, night-time, and I savour the rustle and slap of the Mersey. After the fine grain of the suburbs, it's a relief to stand here and feel space unfold. Over the river, on the Wirral shore, there is a bristle of lights against blackness. I look south, towards Rock Ferry and the *Great Eastern's* last berth, and reflect on the overriding melancholy of her career. She had failed at everything she attempted, and her one success had been an inadvertent failure, or at least had confiscated a ship's most prestigious cargo. But, knowing her story as I now do, I find myself irresistibly warming to this ship with her strange ways, ultimately more loner than liner or line-layer, never quite fitting in anywhere, always set apart from her fellows, regarded from the first almost as an embarrassing mistake. In this way I feel sympathy for her, much as I would for any misfit. Which is strange, for how can warmth or affection fasten to so colossal a thing?

Despite the numerous images of her, and the Saturday mast, and like the American public who believed they had been hoaxed by the 1858 cable, I sometimes wonder if *Great Eastern* ever really existed. Her epic story has a quality of make-believe (though there was certainly the ring of truth in the maddening clang of the wrecking ball against her hull). Perhaps now, as then, it remains difficult to fully comprehend her size. How could something so enormous leave so few traces? How could a ship so vast leave only a mast?

In November 1888, before the arduous work of breaking began, Henry Bath and Sons auctioned off the ship's parts. Everything was for sale, from her metals to her machinery to her interiors to her ephemera. All had to go: they, like so many of her other owners, were desperate for profit. And although the money raised by this auction was later lost in the colossal cost of demolishing her, the auction itself was a success. It drew an unusual mix of buyers, from scrap-metal merchants to souvenir-hunters. Pretty much everything was sold: some of the ironwork went to a Scottish wheelwright, while some was reputedly used to build another Atlantic liner. A Chinese nail-maker bought the wire rigging. And, needing a flagpole for their ground at Anfield, representatives of Everton FC bid successfully for one of the ship's topmasts.*

So, then, it is not only the mast that survives of *Great Eastern*. Gone asunder in the auction, we must imagine pieces of her gradually dispersing, melted and reforged into the nation's fabric, remade as wheels, fenders, hulls, stoves and who knows what else. In this sense, the great ship is not gone, just reincarnated.

---

* Everton FC were the original occupants of Anfield; in 1892, the ground would be occupied by the newly formed Liverpool FC.

16. *Great Eastern*'s mast in its new home at
Anfield Stadium, c.1912.

Propeller

Propeller from RMS Lusitania

LIVERPOOL; HENDON, LONDON
EDWARDIAN

The sky clears to a deep black, a breeze having dispelled the earlier mist, and the stars array themselves like pinpricks. I turn from the river rail and peer into the dock hinterland. The view is of a halogen strip of shops and restaurants on the ground floors of the dark brick warehouses, seen across oily squares of water held within the docks themselves. One of the basins holds the moon, coinlike in its fullness; I have a flicker-memory of an episode of an old kids' programme, *The Night The Moon Came Down To Bathe*.[1]

Somewhere amongst it all is another ship-fragment, perhaps comparable to *Great Eastern*'s mast in potency, but exceeding it in presence. I wander about, pinching the interface of a smartphone map, enjoying the podiatric feel of the cobbles under my thin soles. Were it not for the unfortunate new buildings littering the Merseyside and looming at the docks from the north (Liverpool is one of the most charismatic cities in England, but makes a poor fist of being modern), this would be a nice immersion in the quays as early Victorians knew them, when Liverpool was foremost a city of stuff: tea, spices, silks, fabrics and grain stowed into those commodious warehouses.

Now, however, I'm interested in Liverpool as a city of passengers, its role as jumping-off point for the great ocean liners of the Edwardian age, situated at one end of the trans-atlantic route to New York via Ireland, a passage marked with pathos (in the sorrow of the emigrants) and bathos (in the luxury of the liners). Qualities which, as we have seen, mark the career of *Great Eastern*, and it's satisfying that the last relic of that premature ship should lie in the same city as the last vestige of a prime liner.

On the dockside lies a once-golden propeller, upended

upon a plinth, fanning out its four giant blades. Now sullied with rust, light from nearby lampstands gleams dully on its leading edges. One of a quartet that once drove RMS *Lusitania*, a fast and charismatic Cunard liner of 1907, propellers like this have a disquieting beauty. Ingeniously shaped to tear through the oceans, they reveal just how shipbuilding had then come to focus on superseding wind and wave.

But the propeller looks ill at ease here. After all, despite their beauty, these artefacts were never designed to be seen, or to be still for long. During its operational life, *Lusitania*'s propellers thrummed the ship invisibly towards its ports of call. While passengers admiringly walked the decks, the propellers gleamed out of sight below the waterline. And, strikingly, when redundant they seem to have been valued only for their metals, rather than their aesthetic value. Few were preserved when the liners were scrapped. Uniquely, this one was fished from the ocean floor where *Lusitania* now lies.

Like *Great Eastern's* mast, *Lusitania's* propeller promises thrust: but whereas the mast's simple form speaks readily enough of its invisible fuel, the working of the propeller is more of a mystery at first glance. Those swooping blades, those angles, those curves – these speak of modern science, although they are in fact derived from a form even more primal than the mast: the helix, found everywhere in nature. As it spun, the propeller cut a helical tunnel through the water, developing the thrust which pushed forward the ship.

It's getting late. My mind wanders to the golden ceiling of the Crown, near Lime Street Station, and the lager and steak that await me there. And the more I size up the propeller, the more commonality it seems to have, not with a mast, though they do the same work, but with the Billingsgate Trumpet. What they share is a certain man-made quintessence. Whereas the mast's arboreal origins are obvious, there is something otherworldly about these two strange objects, representing as they do the metalworking of their respective ages carried to the highest pitch. And there is a nice asymmetry to them as a pair. One stands for a tentative, juvenile phase of English seafaring, while the other embodies a late stage in which we briefly considered ourselves unsinkable.

From Merseyside, I travel south to Hendon, London, to the sort of quiet, landlocked street that most Londoners inhabit. On a weekday afternoon I stand in the road opposite a school, the pavement strewn with London flotsam: chicken bones, sheaves of crushed cardboard, discarded cartons. Approximately overlain on my phone are a nineteenth-century Ordnance Survey plan and Google's modern map. I tab between them and pace over the carriageway, trying to approximate my location to that of a lost horse pond.

For it was out of this unlikely body of water that, in the 1830s, the propeller's beginnings were ultimately fished. Old Guttersedge Farm may now be sunk permanently beneath the asphalt and paving of suburban Hendon, but it deserves to be remembered as the place where a farmer discovered how a model ship (and, therefore, a real one) could be driven by an Archimedean screw, a helical device that would eventually develop into the modern propeller.

Francis Pettit Smith was son of a Kentish postmaster. A later portrait of him in the National Maritime Museum depicts an elfin figure with a dreamer's features instead of the ruddy countenance of a farmer. He was born in 1808 near Hythe, one of the ancient Cinque Ports. After private schooling in nearby Ashford, he began sheep farming in Romney Marsh and later moved to Old Guttersedge Farm in Hendon, then part of rural Middlesex.

So far, so bucolic. But Smith was also a lifelong amateur shipwright. He started small, building first to the scale of ponds, then later to that of the sea. Quite from where his obsession derived is not obvious – for example, Hythe harbour had silted up centuries before his birth – but as a boy he showed a rare aptitude for constructing model vessels and a later ingenuity in powering them. At twenty-seven, settled in Hendon with his wife Ann, he had devised one model propelled by a clockwork screw. One day in 1835 he launched it onto one of his horse ponds, watching as it crossed auspiciously to the grassy fringe on the other side.

I imagine this scene unfolding somewhere beneath my feet, and I hanker to know more. Who taught this farmer ship-wrightry? Who introduced him to Archimedes? Whence came his interest in, and knowledge of, hydrodynamics? In fact, the first half of the nineteenth century was a time of febrile invention when it came to ship propulsion. Smith wasn't the only amateur in his field. His Archimedean Screw Propeller (as he patented it) would compete with, and eventually push ahead of, such whimsical mechanisms as Perkins's Sculling Paddle-wheels (mechanised oars attached to the vessel's sides), Nairne's collapsing Propellers (paddles which imitated the motion of a duck's foot) and Linnaker's Propelling Pumps (a jet of water expelled from the vessel).

Of course, Smith was not the first to experiment with screw propulsion. The principle of screw-power had been known for centuries; gondoliers had carved the same sort of arcs through the Venetian lagoon with their oar-blades. Some early propellers were manually operated,* but this was clearly unworkable on a large scale. Mechanised, the helix spiral of the Archimedean screw (an ancient method of water-lifting still in use) was clearly a promising way of driving a ship. There had been many attempts to fit and power a kind of 'spiral oar' (as James Watt termed it[2]) to the rear of a vessel, but the men who dreamt up these ideas either left them on the drawing board or were unable to realise them, usually for want of a watertight method of transmitting reliable, economic power through the hull to the submerged screw.

In 1836, Smith patented his screw-system and exhibited his clockwork model at the Gallery of Practical Science on Adelaide Street, just off the Strand, London. It seems to have been a faintly disreputable place, founded by earnest 'scientific gentlemen' but quickly reduced to an amusement arcade

---

* Such as that of *Turtle*, the curious American submersible built in 1775 and lost in 1776.

('novelty and amusement blended with instruction', according to *The Times*). Here, fascinations of the day were exhibited to a jostling, semi-respectable crowd; Sir John Barrow, a reforming Admiralty official, happened to see Smith's screw and foresaw a use for it in the navy.

In this, Smith was luckier than a rival,* John Ericsson, a Swedish army officer who was simultaneously developing a propeller. Unlike the simplicity of Smith's design, which looked like two turns snapped off a corkscrew, Ericsson had devised an arrangement of two drums on a shaft, each with helical blades facing in opposite directions. Yet whereas Smith was fortunate in being able to deal with a progressive Admiralty official who put him in touch with the steam department to further his ideas, Ericsson had the misfortune to lay his ideas before Sir William Symonds, hidebound Surveyor of the navy (ironically himself an amateur shipwright), who dismissed Ericsson's propeller despite seeing it successfully trialled; Ericsson's ideas live on in the propulsion system of the torpedo.

Presumably with someone else minding Old Guttersedge Farm, Smith threw himself into pitting his screw against the sea proper. At Wapping, in 1836, Smith and Thomas Pilgrim, a jobbing engineer, oversaw the construction of his first boat, *Francis Smith*, a ten-ton launch with a ladle-like hull and a two-turn screw. Although not fast – the engine could achieve only six horsepower – the *Francis Smith* successfully coursed up and down the Paddington Canal and the Thames. On one trial, in February 1837, half the wooden screw broke away. Smith was astonished to find the speed of the launch increasing, so fitted it with a metal, single-turn screw (and accordingly revised his patent). That September, he took the launch to sea, cruising down the south coast to Dover and Folkestone. Here, coastguards favourably reported the dogged passage of *Francis Smith* in an autumn blow.

Encouraged by the Admiralty, Smith formed the Ship

---

* Rivals at first, perhaps, but Smith and Ericsson eventually became friends.

Propeller Company and raised the capital for his first fully fledged ship. Built by Henry Wimshurst of Millwall, *Archimedes* was of 237 tons, fitted with a screw spun by an engine of 80 horsepower (leaving the horse pond far behind). Launched in 1839, *Archimedes* stunned the Admiralty and the engineering world. Here was a full-scale demonstration of the screw as a viable and, indeed, highly desirable method of propulsion. During Thames trials, she made the measured mile at Purfleet in nine minutes, twice the speed that would have satisfied the Admiralty. In May 1840, she far exceeded the Sea Lords' expectations in the Channel, too, keeping up a steady speed of 9 knots regardless of wind or tide.

Onlookers were fascinated by her manoeuvrability compared to the lumbering paddle steamers or windbound sailing vessels. Indeed, she positively sliced through the waters, leaving scarcely more disturbance in her wake than a ship under sail; indeed, she seemed altogether indifferent to the weather. And there were other advantages to the screw-system. For naval purposes, it weighed less and, unlike the paddle-wheels, freed the flanks of a vessel to carry guns; the machinery could be located wholly below the waterline, safe from cannonfire (though not, a little later, from torpedoes).

In summer 1840, *Archimedes* circumnavigated Britain, setting off from the Nore and steaming clockwise to Hull. According to one passenger: 'Smith and his crew deserve to be immortalised. We have beaten everything that came our way – and have passed, maugre [in spite of] the weather, a most delightful time.'[3] By now widely shared, these sentiments contrasted starkly with the scepticism that had hitherto prevailed. Brunel immediately grasped the screw's promise when *Archimedes* steamed into Bristol. After borrowing her from Smith for three months, Brunel radically revised the design of SS *Great Britain* to incorporate a six-blade propeller.

Later that same year, *Archimedes* made the fastest recorded voyage across the Bay of Biscay to Oporto. She then toured Dutch ports. These jaunts were, in effect, an international exhibition of screw-power. Yet even with Brunel's endorsement, uptake was slow. Paddle steamers continued to be built; the Admiralty, though their every expectation of *Archimedes* had been met, did not buy her and they were slow to commission the first screw-driven warship. True, the Sea Lords had in 1840 purchased *Dwarf*, built by John and George Rennie, who after supplying the engines for *Archimedes* immediately built this screw-driven ship. Or there was little *Bee*, completed at Chatham in 1841, which combined screw- and paddle-wheel propulsion. But it wasn't until April 1843 that HMS *Rattler* was launched from the yard at Sheerness.

By now a third-time father (to young Archimedes Petitt Smith), Smith was appointed a consultant by the Admiralty alongside Brunel. For two years they experimented with *Rattler*, trialling no less than twenty-eight different screw geometries, some devised by Smith, the others by different propeller designers. Indeed, the first half of the nineteenth century saw a remarkable profusion of these figures, battling to be recognised in sea-trials and, perhaps most significantly, in the courts as the propeller's rightful parent. Patents proliferated, were revised, were challenged. In the vaults of the Science Museum, London,* survives a group of the scale-model prototypes, and their remarkable variety demonstrates how there was as yet no settled screw design method and no rule of its behaviour; hence the laborious trying-out of all those screws on *Rattler*, to see which pitch, quantity and girth of blades would drive her fastest.

Eventually, Smith's two-bladed design won out, a great advancement on his earlier screws which had more closely resembled parts of a helical spiral. Whereas that which drove *Archimedes* shared much with his earliest patent of 1836 (the model of it reminds me of a fragment of fusilli pasta), in *Rattler*'s propeller can be seen the beginnings of the 'classic' propeller design recognisable today: two brassy blades, angled to cut swathes through the water, springing from a central boss.

---

* Collected by Bennet Woodcroft, one of Smith's rival screw-designers, and bequeathed to a forerunner of the Science Museum.

No doubt Smith learned much from his competitors, many of whom had proposed variants on this type.

Despite these extensive tests, the Sea Lords required further convincing before they would turn the navy over to the screw. So, a contest was devised between *Rattler* and HMS *Alecto*, a paddle steamer of similar hull design, approximately equal tonnage but slightly inferior horsepower. Their wooden hulls were designed like those of sailing ships, and indeed both possessed a full complement of masts, the navy at this time viewing steam as a standby for, rather than displacement of, wind power: the two warships therefore rode awkwardly between the two ages. In March 1845 they raced solely by steam, or solely by sail, and under both. *Rattler* consistently outperformed her opponent. Finally, both ships were lashed together for a stern-to-stern tug-of-war. With both their engines working hard (for this was no longer just a speed-trial, but now a matter of prestige), the paddle-blades splashed the surface and the screw whumped furiously below. *Alecto* began to drag *Rattler* backwards, before the latter hit the pitch of her engines and heaved *Alecto* decisively astern.

In the 1850s, having played a pivotal role in convincing the Admiralty of the screw's viability, Smith found himself fighting with other screw-makers in the courts for official recognition. Later defending himself in print, he acknowledged that, 'The idea of using the screw as a propeller is at least a century old, and within that period has had its hundred votaries, each in turn nursing it as the offspring of his own fertile imagination, until compelled, by failure of repeated trials, to abandon it in disgust . . .'[4]

But some of these votaries had made effective refinements to the screw. Robert Griffiths, for instance, realised in 1849 that the central boss of a propeller could be enlarged to give it extra strength without any attendant deficiencies in thrust. His patent of that year came to form the basis for a 'classic' propeller design in widespread use in the mid-nineteenth century; in 1860, he would leave for posterity a pamphlet wistfully titled *The Screw Propeller: What it is and what it ought to be*. And there was John Penn, a marine engineer who in 1854 finally solved the problem of water ingress through the propeller shaft socket by lining it with lignum vitae, a self-lubricating form of wood.

There was also Henrietta Vansittart, child of James Lowe, one of Smith's contemporary competitors, who would doggedly patent and promote several novel screw designs (including one trialled on *Rattler*) and in doing so bankrupt himself. Henrietta had become fascinated by her father's work, joining him aboard HMS *Bullfinch* for trials of one of his designs in the 1850s, and after his death continued to develop designs for a screw. Eventually, in 1871, her patented Vansittart-Lowe propeller would be awarded a first-class diploma at the Kensington exhibition and fitted to numerous vessels, including a namesake of the *Lusitania*.

What drew all these votaries to the propeller when not a few of them, like Lowe, were ruined without recompense? I think the reason is twofold. Obviously, the prospect of wealth and fame was significant – after all, the propeller promised to upend seafaring forever. But, at the same time, the propeller was itself a bewitching artefact – a form which seemed capable of endless variations, with the promise of the 'perfect' shape forever beckoning at the end of another day's tinkering. (It transpires that, even today, there is no Ur-shape for the propeller, no single solution applicable to every stern, or perfect form that would beat the sea's friction every time. Indeed, there is still much that is unknown about the behaviour of propellers in the complex currents of water at varying sterns in varying oceans.)

Despite fierce competition, Smith can, I think, claim the propeller as his own: he did more than anyone to prove it, first in model form and then in life-size demonstrations, as a tool by which to master the sea. After much legal wrangling, the screwmakers eventually settled for a shared payment of £20,000 from the Admiralty. For Smith, however, this was scant compensation for the Sea Lords' decision not to purchase *Archimedes* (as he had been given to understand they would if she proved a success). From this economic blow his company and his marine ventures reeled, and then sank upon the expiration of his screw patent in 1856.

By then, he had also lost little Archimedes, who in 1848 had died at the age of seven and was buried in the family plot at St Leonard's Church, Hythe; *Archimedes* had meanwhile become a sailing ship, her screw and engines having been ignobly removed, and Smith had retired to Guernsey to farm once again.

Not far from the site of Old Guttersedge Farm, following a bridge over the roaring abyss of the M1, lies the Welsh Harp reservoir. Nicknamed from a lost tavern nearby (and officially named the Brent Reservoir), it was dug in the early nineteenth century to feed the water levels of the Grand Union Canal, but is now a tranquil haven for wildfowl. I stand at the water's edge, light lowering over Hendon, watching a chevron of geese arrow into the sunset.

As Smith farmed disconsolately in Guernsey, ships were crisscrossing the Atlantic in swarms – or lines. Under sail, this had been a wayward and unpredictable voyage to a British colony and then embryonic post-colonial nation, with no guaranteed scheduling of arrivals and departures. But with steam came greater certainty in ship movements – ably demonstrated in the 1830s by Brunel's *Great Western* – and by this time America was a new nation, young, garrulous and eager to make its mark upon the world.

Paddle steamers began plying between America and Great Britain, operated by a cohort of companies that offered fixed departure and arrival dates. Most of these are now long extinguished, but two names have endured: the White Star Line and Cunard. Although the fortunes of some were underpinned by government mail contracts, competition between these 'lines' was ferocious, speed fetishised. Spurred on by the way steam unlocked unprecedentedly short Atlantic crossings, these liners extracted all the speed from their paddle wheels, augmented by sail, that it was possible to obtain. But there was a ceiling to paddle-wheel speed.

Fortuitously, the screw found widespread acceptance at just the right moment. It suited passenger liners as much as it did warships, and for the same reasons: lighter machinery and unencumbered decks (this time for sweeping views of the sea, rather than for sweeping gunfire). As it had been with the dockyards of the Admiralty, it was some time before the propeller crept into commercial shipyards as a trusted source of motive power; but by the 1870s the screw had displaced the paddle-wheel. It drove the liners much faster and led to races between them to wrest away the Blue Riband, as the record came to be known – the unofficial laurel for the quickest transatlantic crossing.

In the wake of Brunel's pioneering *Great Britain*, new, single-screw liners began to appear upon the Atlantic. First off the slipways, in 1850, was *City of Glasgow* for the Inman Line. Today, this line is semi-forgotten, but it was the first to invest in a propeller-driven fleet for transatlantic crossings. Others followed.

In 1871, RMS *Oceanic* made her maiden voyage from Liverpool to New York. She was one of a family of sleek new screw-liners commissioned by the White Star Line; her sister, *Adriatic*, would be the first screw-driven ship to win the Blue Riband. In 1871, also, Smith was knighted in recognition of his role as propeller-pioneer. By this time, friends had lobbied him out of obscurity, securing for him numerous honoraria including the curatorship of the fledgling patent office museum; and when, in 1874, he died in the genteel surroundings of South Kensington, the screw he had tenaciously promoted was in widespread use.

In the closing decades of the nineteenth century, British, American, German and French shipping lines were racing one another across the Atlantic for the Blue Riband in a new breed of vessel, streamlined in tanks before they rose on the slipways. Large, fast and with prominent smokestacks (and tokenistic masts), ships like White Star's *Teutonic* (1889), Cunard's *Campania* (1893) or North German Lloyd's *Kaiser Wilhelm der Grosse* (1898) were realising the promise of their ultimate ancestor, Brunel's *Great Eastern*. In 1899, a second RMS *Oceanic* – a daughter-ship, perhaps – was launched for the White Star Line. She was the first to outdo Brunel's Great Babe in length, if not tonnage; below her stern spun two propellers, each failsafe for the other, together achieving undreamed-of thrust.

Their propellers were driven by variants of the reciprocating steam engine, a thing of cumbersome moving parts that shared more in spirit with the paddle-wheel than the elegantly spinning propeller. But the propeller would soon find its soulmate.

On 26 June 1897, the Diamond Jubilee review of the Royal Navy was held at Spithead. It was a pageant of naval power. Aloft, countless lines of flags and pennants fluttered excitedly; afloat, four five-mile lines of warships – 166 in number – kept formation in the sheltered waters of the Solent.

With the seventy-eight-year-old Queen too frail to attend, the Prince of Wales (later Edward VII) presided on her behalf. No other country could equal this display; no two countries, even, could join their naval forces to outnumber the warships riding in that thirty-mile tract of sea (and it was not even the entire fleet, some still being stationed in the Empire's far corners). There were many onlookers, from spectators thronging the shore to 'spectatorial craft of all kinds . . . from the snowy-winged pleasure yacht, skimming over the waves like a seagull . . . to the mighty-hulled ministers of commerce and intercourse called liners, which even seemed to dwarf some of our biggest battleships . . .'[5]

These 'spectatorial craft' kept respectfully to the edges. Indeed, it was as much an exhibition of late imperial values as it was of the vessels of that age: the faultless etiquette of the people, the rigid discipline of the ships. Salutes boomed and rolled over the waters, while the Royal Yacht moved at a stately pace between the sterns. Everyone knew their place.

Everyone, that is, except *Turbinia*. Uninvited, un-announced, unknown in fact, this experimental proto-speedboat made a mockery of the display. Stunning the onlookers present – who, landsmen or seamen, had never seen anything so fast – she burst into the massed fleet and capered about the solemn lines of warships at 34 knots, an incredible speed, then unmatched by any of the ships present. The guard-ships sent to catch her trailed humiliatingly in her wake. Then she absconded.

Steel-built at Wallsend in 1894, narrow-hulled *Turbinia* was the brainchild of engineer Charles Parsons, who had devised her as an 'ocean greyhound' to demonstrate his steam turbine engine. From the Latin term for spinning top, a turbine is an assemblage of slender blade-fringed wheels through which steam passes, catching the blades, spinning the variously sized wheels and generating power; much more power, in fact, than the sort of reciprocating steam engine used in ships to this point.

After first producing turbines for power plants, Parsons turned to marine propulsion. For *Turbinia* he coupled a turbine to a propeller but, initially, the results were unimpressive: 19 knots in early trials, when her 1,000-horsepower turbine was supposed to provide 30 knots. Eventually, Parsons and his team discovered that a faster propeller does not equal a faster vessel. Spun too fast, the propeller carves a cavity of air within the water, *slowing* the speed of the vessel by temporarily removing the fluid upon which the blades act, as well as damaging them.

Ahead of the Spithead Review in 1897, *Turbinia* had been fine-tooled, her horsepower spread between three turbines, each driving a shaft threaded with three propellers. Of course, her appearance there was a publicity stunt, but it was an understandable move, given the conservatism of the navy to innovations when promoted through official channels. Most probably Parsons would have known of the scepticism attending Smith and Ericsson when they had promoted their propeller designs; and Parsons could be sure, that day of the Review, that the eyes of the world would be upon Spithead.

In effect, he made the Admiralty an offer they couldn't refuse. And the turbine promised unprecedented speed not just for warships but also for liners to outmanoeuvre their rivals. Thereafter, as if to demonstrate how highly advanced Britain's shipbuilding industry was by this time, it took just a decade before turbines were fitted to a game-changing battleship, HMS *Dreadnought* (1905), and a fabulous superliner, RMS *Lusitania* (1907).

Ice cracks and slips, with a rattle, a half-cube's depth down my glass. In the empty dining room of the Hotel Russell, I hear the echoing pacing of receptionists and the clatter of crockery as waiting staff plate up for the day. I chose this lull, the doldrums between meals, to drink in the details of this opulent space. Like the massive hotel in which it sits, the room is adorned with the decadent architecture of the French Renaissance; the room, like the hotel, enshrines Britain's character and tastes at a late stage. How to describe it? Adjectives like grandiose, confident and limitless propose themselves. Unsinkable. I sip my drink.

Built between 1892 and 1899, the hotel was designed by Charles Fitzroy Doll, hotel specialist and estate architect to the Duke of Bedford. Imperious in scale, imperial in architectural language (dominated by the aforementioned French Renaissance, but within a bewildering meld of periods, styles, stones and metals), the hotel, like others of this date and type, embodies a nation that saw itself at the forefront of the world; a nation which, inch-by-inch, century-by-century, had extended its reach to the furthest continents and furthest seas, had assumed control of them as far as it was able, and had yoked these places and those in between back to itself. In London, the epicentre, were built palatial offices and hotels like the Hotel Russell which, today, commemorate this late-imperial phase of Britain.

They also monumentalise their lost counterparts: the ocean liners. For, from the late nineteenth century, the race for speed had been paralleled by growing competition in luxury,

in facilities, in accommodation. Speed could be extremely uncomfortable: the sacrifice of hull space for engines, the pitch
at which those engines worked causing parts of the ship to vi-
brate horribly (one German liner so afflicted, the *Deutschland*,
was nicknamed the 'Cocktail Shaker'). So, the focus shifted to
attracting passengers not with speed but instead with opulent interiors, sumptuous dining, dazzling entertainment: the
amenities of the land.

It was with the next generation of ocean liners, built
or ordered in the first decades of the twentieth century, that
their famous character of hotels afloat came to the fore; and,
meeting them halfway, as it were, hotels increasingly came to
resemble liners ashore. Fitzroy Doll is also known for designing interiors aboard the *Titanic*, including the first-class dining
room. Supposedly it was a reworked form of the one he provided for the Hotel Russell, in which I now sit. I don't know how
alike the two were and, in a sense, it doesn't really matter. Both
interiors came from the same hand and held the same luxuri-
ousness at their core.

Above me, a grid of gilt and stucco, borne on paired Jacobean columns, strides to the vanishing point. For this room is
large, too large in fact to see clearly to the other end from my
chair near the entrance. Beyond is an arrangement of empty
tables, perfectly laid and still, a tableau of linen and silverware.
I finish my drink, then pad out of the room on the soft, deep
carpet, through an extraordinary sequence of empty staircases,
vestibules, corridors silver-leavened and lined with undulating
tones of marble; I am padding, it afterwards occurs to me, as
through a deep-seascape.

Lusitania was a province of the Roman Empire, roughly corresponding to modern-day Portugal and a little of western Spain.
Just who the Lusitani were is still unclear, but it is possible that
they were originally Celts. Like us, they were Atlantic-facing; like
us, they were visited by Romans in the late years of the Republic,
then annexed by Rome in the early years of the Empire. However,
with two millennia between its conquest by a foreign empire and

its own acquisition of one, late-imperial Britain sought kinship with the empire-builders rather than the annexed.

Cunard's naming of *Lusitania* and her sister *Mauretania* (and their third, later sister *Aquitania*) went beyond board-room whimsy. When they were ordered, Britain faced losing its transatlantic pre-eminence. In the early years of the twentieth century, Cunard had become the sole British transatlantic liner operator, the White Star Line and others having been absorbed into a powerful American combine: International Mercantile Marine. Liners owned by the Americans – and the Germans – were outstripping Cunard's fleet and sequestering the Blue Riband. Cunard appealed to the British government for a subsidy to build two unprecedented superliners that would win back pole position in the Atlantic.

These sisters would reassert Britain's ocean primacy. Built partly with public funds, their names sought to project a nation's sense of itself. Lines had received government subsidy before, of course, in the form of mail contracts, but this was something different: a direct loan of monies at generous rates of interest. The government agreed to put up the funds on the proviso that the liners could mount guns in wartime and manoeuvre to the Admiralty's satisfaction.

First to be laid down in 1904 was *Lusitania*, in John Brown & Co's shipyard on the Clyde. She and her sisters were designed by Leonard Peskett, an amiable shipwright who became an authority on liner design. Apprenticed to a Rye shipbuilder (the seagoing of this ancient, silted Cinque Port was not yet over), he spent four years as a naval draughtsman at Chatham Dockyard before being appointed draughtsman at Cunard, where he rose to the uppermost post of Naval Architect. He joined in 1884, when the line's last single-screw liners, *Etruria* and *Umbria*, were being completed; he had a hand in them, and in all Cunard's subsequent liners, from the first double-screw expresses to the slower, steadier emigrant ships. Each ship leaving the slips saw Peskett further perfect his art.

From an early model built by Peskett, *Lusitania*'s hull-form was subject to heavy testing, first in the Admiralty's

Haslar tank and then in that of John Brown on Clydeside. In these tanks, Cunard wished to find the optimum combination of hull, propellers and engines to reach the holy 25 knots that would sink their competitors; from them emerged a hull that reminds me of one of those fast, streamlined modern locomotives, every line drawn towards a vanishing point.

Submerged at her stern were a quartet of propellers, each incorporating three angled, spoon-like blades around a spiky, conical boss, each cast from phosphor bronze and measuring seventeen feet across. The two inner propellers were mounted either side of the rudder, spinning inwards, while the outer two were mounted further back in the hull, spinning outwards; of each pair, one turned around the clock and the other spun the clock back. In a 1906 photograph of *Lusitania* on the stocks, half-built up to the waterline, they glimmer like jewellery against her dark hull.

Each propeller capped a long shaft reaching back into the depths of the coal-crammed hold. There, four turbines turned these propeller-shafts with a cumulative 68,000 horsepower, fed by four groups of five boilers incorporating 192 furnaces. When going full tilt, these devoured 840 tonnes of coal per day, flung into them by hundreds of stokers; they vented their smoke through a series of ducts converging under four colossal funnels, oval-shaped and angled rakishly back from the perpendicular.

Such was the vehicular character of the ship, most of which lay hidden below in the depths of her steel hull. But there was a dual nature to *Lusitania*, as there would be to her contemporaries and to her descendants. Above the waterline, she was a palatial building, her hull honeycombed into hundreds of elaborate rooms, corridors, stairways and promenades. Apart from some third-class cabins located deep in the bows on the same deck as some of the engine machinery, and the fact that the funnels rose through the first- and second-class areas, *Lusitania*'s interiors were terrestrial, not marine.

Not that she was the first ship to incorporate lavish interiors. Larger sailing vessels had often incorporated pleasant quarters ornamented with simple architectural motifs - columns, window surrounds - bent to the ship's lines. But, on the whole, passengers and crew alike were expected to adapt to the cramped warren of oddly shaped rooms, shrunken passageways and head-hitting beams that comprised the interior of a masted ship, and to endure these nautical proportions for as long as it took to reach land.

Iron had freed ships to grow larger and, with more internal space, it became possible to incorporate rooms set out on cubes instead of curves. Suddenly, a semblance of landed normality was possible. Rooms could be orthogonal, unaffected by the hull shape, with headroom and potential for decorative schemes. Again, Brunel's *Great Britain* and *Great Eastern* led the way. But until the closing decades of the nineteenth century, designed, comfortable ship's interiors were exceptional, even on the transatlantic passenger ships. Charles Dickens was famously aghast at the miserably confined cabin he was to inhabit on a fortnight's crossing to Boston in 1842.

True, *Lusitania* had chiefly been designed to outrun her competitors, but it was inconceivable that she should do so without style. The interiors of her competitors had been designed by architects since the closing decades of the nineteenth century. Mewes and Davis, for instance, designed the interiors for the Hamburg America line, while White Star engaged Richard Norman Shaw as consultant. P&O worked with the Arts-and-Crafts architects T. E. Collcutt and J. J. Stevenson. And, of course, Fitzroy Doll devised *Titanic*'s interior. Now Cunard, too, employed an architect to design her principal interiors.

In some liners, bombastic imperial styles prevailed, while in others there was a little more restraint; all, however, replicated the aristocratic architectures of the land. James Miller, a Glaswegian architect with a flair for railway stations and hotels, provided *Lusitania* with sumptuous suites of gilt, plasterwork and stucco, in a subtly varying combination of styles: the neoclassical first-class dining salon was frothily frescoed, while the more Palladian lounge was panelled with restraint. Miller's execution of these styles was tasteful, stopping short of the crassness risked when mashing together differing aesthetics. German liners, in particular, could be very overripe.

In all liners, second- and third-class interiors were more muted, their design and execution usually left in the hands of the shipyard. Reflecting society on land, there was of course strict segregation between the various classes of passenger, expressed in the decor and physically delineated in the arrangement of the ship. *Lusitania,* however, was notable for the quality of her third-class accommodation, which even incorporated a small external area for the (mostly) emigrant passengers to share the ocean air; and anyway, thanks to the propellers, all parties shared in the ship's speed.

Whatever the style or intensity of embellishment, the first-class quarters were indistinguishable from rooms to be found in hotels, gentlemen's clubs or, indeed, stately homes; the more modest second- and third-class accommodation, too, resembled its terrestrial counterparts. And what it all amounted to was an unprecedented shutting-out of the sea from the ship. Inside, about the only concessions to its presence were bolted-down chairs and chandeliers reinforced to prevent them swaying excessively – although with the great improvements in hull design, these enormous vessels were far more stable: only the occasional *whump* of the bow breaking an awkward wave was heard and felt in the interior. For most passengers the faint tremble of the engine quickly receded from notice.

Unfelt in the ship's interior, the sea was repositioned as a commodity to be consumed from designated areas of the deck. Uncluttered promenades allowed for first- and second-class passengers to exercise and to view the ocean. From up there, forty feet above the waterline, the Atlantic was all-encompassing, yet held at a comfortable remove, an intriguing panorama rather than a visceral presence. Closer perspectives could be obtained through the portholes studding the hull, which offered greater details of the ocean's heaving surface kept securely behind inches of brass-rimmed glass (though these could also be opened to let in the breeze).

On her maiden voyage in 1907, with her powerful new

turbines and stately quartet of screws, *Lusitania* easily won the Blue Riband back for Britain. With her sister *Mauretania* she squashed the distance between Liverpool and New York from months to just four days. With their architectural interiors in the latest tastes, these ships also shrank the cultural distance between the sea and the land. No longer did passengers have to bend themselves to the ship's lines. Liners like *Lusitania* allowed us to impose ourselves upon the sea, to set forth upon it on our own terms, to no longer tolerate the limitations in comfort or headway that it had previously dictated.

Of course, no passenger really forgot they were at sea – the effect of *Lusitania's* palatial interior was more like a 'pleasurable hallucination'.[6] But just as the cultural distance between sea and land narrowed, so the gulf widened between the passengers and the water. Experienced from a superliner, the Atlantic was abstracted like never before, its cold physicality passed over. Immersed in the opulence of the interior, a first-class passenger might pass the whole voyage without getting wet, or seasick, or fearful of the abyss below. And were I to be bow-tied, three courses through an Edwardian feast, claret in my glass and Debussy tinkling in my ears, then I too might forget about the sea for a while.

Back in Liverpool, I revisit *Lusitania's* surviving screw. It's daylight this time, and the sad, rusted state of the blades and boss is plain. On this chilly September morning, they seem neither to hold nor reflect the bright sun over the Mersey, though I remember how they had gleamed sullenly under the lamps. But I see clearly now, in this light, a whisper of the Archimedean helix in the pitch of the blades, the principle brought to perfection through the testing tanks; I see the link between this kingly artefact and Smith's humble screw, and I see the bond between this majestic port city, where *Lusitania* was refitted with four propellers like this one, and the nondescript London suburb where the screw arose from the horse pond.

Lusitania's surviving propeller lies in the same sidelong way, though there is no gold against snow here, only rust

17. First class dining saloon,
Lusitania. Photograph of c.1905–7
showing the tables laid

against efflorescence, white salt crystals blooming below the blades and spreading across the pitch as if in suggestion of snow. Rust, salt, decay – it's clear the propeller has no business being exposed to the air like this, especially so near the coast. Phosphor or manganese bronze were used for propellers because they proved most resistant to marine wear; the resultant beauty was happenstance. But even these durable metals don't last long when left outside.

Lusitania and her sister set the general tone for liner design over subsequent decades. They were followed by the trio of superliners commissioned by Cunard's chief competitor, the White Star Line: *Olympic* (1911), *Titanic* (1912) and *Britannic* (1915). Surpassing all others in their size and luxury, these vessels were built for comfort rather than for speed. In addition to their vast tracts of sumptuous architecture, swooping staircases and profusion of dining options, these ships incorporated such things as gyms, swimming pools, Turkish baths and squash courts. Did the capability to do this breed a degree of complacency towards the sea, of hubris even?

As with iron, steam and screws, marine engineers had long quested after an 'unsinkable' ship design. In 1838, in Paris, a boat built around two watertight metal barrels was floated in the Seine and was claimed to be unsinkable – though, as a contemporary newspaper reported, the idea was 'anything but new'.[7] But by far the most promising step was the compartmentalisation of the hull. Bulkheads created a cellular interior so that if the hull were holed, the ship should remain afloat, if not upright. Originating in Chinese shipbuilding, bulkheads first appeared in Western vessels in the paddle steamer *Garryowen*, built in 1834 to ply the Lower Shannon.[8] Subsequently they found widespread adoption. Samuel Plimsoll, when viewing the drawings of the Inman Line's new *City of Paris* in 1890 – the hull of which consisted of thirty individual compartments – opined that such a vessel would be 'practically unsinkable'.[9] *Lusitania*'s hull was transversely divided into twelve watertight compartments; *Titanic*'s into sixteen.

It is often said that the paucity of lifeboats provided aboard liners illustrates a misplaced confidence in their hull design and a resultant hubris on the part of shipowners, passengers and the public at large (though not, I must emphasise, amongst sailors). Actually, government regulations of the time were satisfied with so small a number, the growth of these huge ships having outstripped the law. More indicative of hubris, I think, was the desire to equal the scale of the ocean itself in force and tonnage, the desire to align the ships' interiors with the best terrestrial buildings (hotels don't sink) and the desire to exclude the sea from the ship or, at least, to neuter its presence: these things led to a kind of oceanic complacency.

A May morning in 1915 found the *Lusitania* steaming under the southernmost tip of Ireland. She had run homeward through a much quieter Atlantic, almost the only superliner still in passenger service in the early years of the First World War. Germany, stepping up hostilities, had declared British seas a warzone and its U-boats were attacking both naval and civilian vessels; by this time, the torpedo had become a viable naval weapon. Passengers and crew aboard *Lusitania* had been warned by the German Embassy in Washington, prior to embarkation, that 'vessels flying the flag of Great Britain . . . are liable to destruction in those waters . . .' Nevertheless, she sailed, knowing the U-boat threat.

A little after 2 p.m., on 7 May 1915, *Lusitania* was torpedoed by *U-20*, which had already sunk three vessels in this danger zone. The torpedo hit the ship to starboard, near the bridge; a vast explosion sundered the hull, quickly followed by a second, the cause of which, to this day, is still not fully understood; the German justification for attack was that she carried munitions, which was in fact true, although it amounted merely to cases of rifle ammunition stowed deep in the hold. She also carried 1,960 men, women and children. Only 767 of them would survive. Quickly she heeled to starboard, immersing her fine prow, and a few minutes later the lifeboats were chaotically launched. Panic reigned, watched by *U-20*'s captain through his periscope.

Seawater crept quickly through the ship. It lapped under doors and poured through keyholes. It swallowed the skirting and felt for the dado rails. It unpeeled the wallpaper and dissolved the gilt. Perversely, it filled the baths, the sinks, the sanitary ware; filling room after room, it caressed the plasterwork ceilings. Flooded thus, the urbane interiors became nightmarish; for so long excluded, the sea had made its presence felt inside. Outside, as the ship tipped forward, her stern rose to expose her golden propellers under the bright sunshine. Momentarily they spun in air, not water, before *Lusitania* finally plunged down.

Over sixty years later, this propeller was raised and brought ashore, then bought from the salvors by Liverpool Maritime Museum and placed here (two others were salvaged at the same time and went to America). And I find myself thinking again of the Billingsgate Trumpet. Despite the catastrophe, despite the ravages of the open air, the superb cast of the metalwork impresses still.

It's not one of her original three-bladed propellers, but one of the four-bladed replacements fitted in a Mersey dry dock in spring 1909. The outermost were sourced from John Brown & Co., while the innermost were manufactured by the Manganese Bronze & Brass Co. on the Thames. Each demonstrated Britain's supremacy in metalworking, and cost the equivalent of a quarter of a million pounds.[10]

It's a many-bladed monument, this propeller, one that poignantly commemorates a tragedy but also beautifully illustrates a mindset. It embodies a time when our ingenuity – or so we thought – could override the sea's force. Unrivalled catastrophes such as the sinking of *Titanic* and *Lusitania* checked this belief. They resulted in safer ship design, improved regulation – greater humility, in other words, towards the ocean. Cruise ships, the superliners' modern successors, may still resemble floating settlements, but they will always be shadowed by their lost predecessors.

Mercifully, losses of ocean liners were rare. Most of them were scrapped when they had been superseded in thrust and luxury by successors. And because they were so well known to millions – not least through canny advertising – their final voyages to the breakers' yards were always deemed worth seeing. *Lusitania*'s sister, *Mauretania*, outlived her by a good twenty years, retaining the Blue Riband for much of her life, and escaped the First World War relatively unscathed. In 1934, she was retired from service and broken up at Rosyth.

Shortly beforehand, her interiors were sold at auction. Mr Cullen, a Lloyd's underwriter, bought the second-class Drawing Room and installed it in his new house overlooking Poole Harbour. Ronald Avery, a Bristolian wine merchant, purchased the mahogany columns, panelling, mirrors and dome of the Library and installed them in 'the first of Avery's New "Mauretania" Bar Lounges (Lower Deck!)'[11] on Park Street, Bristol. Other rooms went to a Guernsey hotel and Sunnyside, a little house on the Isle of Wight. And this fragmentary afterlife wasn't confined to *Mauretania*. The owner of the White Swan Hotel, Alnwick, snapped up for an extension the dining room from RMS *Olympic*, scrapped in Rosyth at the same time as *Mauretania*; panelling from both ships adorns the eastward parts of the Catholic church of St John the Baptist, Padiham, Lincolnshire.

While certainly impressive, these re-established liner interiors are more prosaic than those formed by reused ships' timbers. After all, the rooms were conventional architecture which simply happened to float for a time, rather than being naval architecture mysteriously incorporated into houses, pubs and churches. Life within them was life on land, wholly free of the hardships the ships' timbers had seen. Civilisation, these reused interiors seem to suggest, was demountable.

Liner propellers seem never to have been deliberately saved. Being submerged in their active lifetimes, their profile was far lower to those passengers who fell in love with their ships. Practically, at least, there were also certain obstacles to their reuse: these were vast, heavy objects, and it would have been costly to outbid the scrap merchants to whom they must have represented a bonanza. Yet I lament the fact that so few survive. Unlike the terrestrial interiors they drove through the oceans, propellers were genuinely nautical, emblems of our greatest confrontation with the forces of the sea. And in the propellers designed for the liners, they reached a pitch of perfection and prestige never since attained.

I circle the screw once more. For the majority of its life it spun vertically, pointing back to the departing land. Now, it lies incongruously sidelong, boss pointing skywards. It occurs to me that, like the Billingsgate Trumpet, the propeller speaks of the sea but is not solely nautical. Another branch of its ancestry produced props that bit air, rather than water, and drove the first aeroplanes. And, serendipitously, it would only take about three-quarters of an hour to walk across Hendon from Francis Smith's old farm to a nearby aerodrome where commercial aviation was pioneered; perhaps less if it were straight across the fields Smith knew.

Every so often you come across an image in which the shifting of ages is caught. J. M. W. Turner's *Fighting Temeraire* is perhaps the most famous, but there are others which are just as compelling. For artists, depicting a pivotal moment or innovation has obvious attractions, promising as it does to provoke in the viewer both nostalgia for the old and shock at the new. Another fine example is Eduardo de Martino's painting of the 1897 Fleet Review at Spithead, showing *Turbinia* frolicking amongst the warships.[12] There are also the iconic photographs taken in New York in 1909, which show the Wright Brothers' revolutionary new Flyer at a standstill in a field overlooking the city. Passing behind is *Lusitania* as she makes her entrance to the harbour, Statue of Liberty in the distance.[13] Despite her status as the most sophisticated vessel then afloat, she occupies the background of the photograph; as if the scene were snapped by a soothsayer, the plane stands in the foreground, ready to supersede her.

18. *Lusitania*'s propeller

Hull

Hull pieces (various) from Rosebud

NEWLYN, CORNWALL
TWENTIETH CENTURY

Her features can just about be discerned amidst a pile of cur-vaceous timbers on Dynamite Quay. For years she had lain hulked at her moorings, ruptured by raindrops, worse for ships' timbers than brine. Now her hull is a hell of planks.

From this quay, it's only a three-hour tramp across rugged country to the sandy beach where she was built, close to a century ago. And during her life she had not strayed all that far, except for one dazzling voyage away from her fishing-grounds. Now she lies in fragments, brought in from the coils of the estuary where she had been left.

It won't be long before she disappears for good. But word starts to spread. Soon a long file of people assembles, each to carry off a piece of her carcass. *Rosebud*, once so iconic a fishing vessel, has become another kind of fishing-ground.

My wife drops me next to the public conveniences and motors off to Land's End before Ida has a chance to wake up. Out of the car, I'm hit by a cold south-westerly skipping off Mount's Bay. I pull my cap down low on my forehead and shove my hands in my pockets. Wearing just a T-shirt and overalls, a purple-haired fisherman strides purposefully out of the Co-op with a flagon of cider. He smiles at me, I smile back. Hello Newlyn, I murmur to myself, once he's gone around a corner. Twelve bells, says the town clock, and I climb up Chywoone Hill to-wards the Fradgan, an old haunt of fishermen.

Up, up, up the hill, good exercise for the calves, which I need after indulging in all those fish suppers and fresh beer. I climb up towards a cloud-tumbling sky, passing respectable houses soberly lining the hill. Up, and then at a steep junction

I see the way into the Fradgan,* under a whitewashed house set back from the others, breaking their carefully observed building line, its older age betrayed by its illogical frontage and the narrow passageway it offers into the backstreets.

Through the passage and onto Orchard Place, backbone of the Fradgan. Here, low granite cottages huddle in shoals of cobbles. An asphalt path straggles downhill, glancing off walls and out of sight somewhere below. I dawdle down this narrow way. At first, the Fradgan reminds me strongly of a reef: rubble walls dripping and glistening in the rain, ruffled with green plants waving on windowsills, riddled with crevices and passageways. Reefs may be death to ships, but they brim with life of their own. So it once was here.

Little dead-end lanes stem from Orchard Place like fishbones. One is named Chapel Street; others seem to be nameless. The cottages crowding them are remarkable: built of irregular granite blocks, some huge, others tiny, mortared together to form dwellings that no storm could ever defeat. In them once lived generations of fishermen and their families. But now, apart from a couple of cats, the Fradgan seems lifeless. Nothing stirs but the wind and the plants, not even in the windows. Perhaps this bullying south-westerly, raging about the place like a trapped bull, keeps everyone inside. Or perhaps there's a more fundamental reason.

At its bottom, Orchard Place narrows almost to a slot, excluding it from Google's cameras, which have usefully though joylessly trawled up the world. I squeeze through to join The Fradgan, the crooked street from which the whole tract takes its name. Here I am stopped in my tracks by the irregularity of its building line: the cluster of cottages on the south side of this street juts forward and back from it to a remarkable degree, to the extent that some of the dwellings actually face different bearings of the compass. There is a pleasing absence of logic to this, as if they obeyed no pattern or rule other than their own caprice. In fact, they are like boats frozen mid-jostle at the

* Pronounced 'Frajan'.

harbour wall, which is appropriate since they once backed onto a real harbour called Keel Alley; boats once bobbed at these housebacks.

Half-hour chimes filter into the Fradgan. I must keep an appointment at a boathouse with pale blue doors some way to the west of here. I should get going. But something about this place roots me. I draw the comparison with a reef again. Like a reef, the Fradgan and its fellow ancient areas of Newlyn are irregular, organic, unsquared. Like a reef they were hives of activity, families crammed into these small cottages, turned in and out by the tides, bundling through the narrow ways with nets to be dried and mended, pilchards to be salted, packed and pressed, or shelter from a gale to be won.

And like all reefs they pose a threat to ships – of state. In the run-up to the Second World War, the Fradgan, the Narrows, the Navy Inn and other ancient parts of Newlyn faced an existential threat from the local authority. But its schemes were defeated by a vessel crewed by men from these ancient quarters, a vessel whose name will forever be currency here: *Rosebud*.

I leave by Gwavas Quay, a former slipway which now connects the Fradgan with the Strand, the town's main highway. Like so many Cornish place names – 'Chywoone', 'Fradgan' – 'Gwavas' carries the sense of foreignness that Cornwall possesses. Gwavas Lake, as this north-westerly corner of Mount's Bay is known, was the best place to shelter from the vicious southwesterlies off the Channel. It's aptly named. Cornish 'gwavos' means 'winter farm' or 'winter abode'.

The wind blows more fiercely on the roomier Strand, filling my coat as if it were sailcloth. Overlarge lorries roar past, having squeezed through the pinched, medieval roads further along at Tolcarne. Incongruous though they seem, they are here because this fishery continues to prosper. Incomers to Newlyn may be surprised by how greatly the town's quayside differs from those of other Cornish harbours, those which are perfectly preserved as if in aspic. Here, tourists do not have the run of the place. Newlyn continues to be a working port, views of its quays blocked by utilitarian sheds. It has not succumbed to its own picturesqueness like the others – which gives it less surface appeal, perhaps, but deeper charisma.

Newlyn's little quay enters the historical record in 1435. It was rebuilt that year with indulgences – or remittance of sins – promised to the donors, and has been much rebuilt since. As I hurry down the Strand, ungainly modern buildings give way to a view of trawlers at the old quay. For over four centuries, this little breakwater withstood the gales which delve into Mount's Bay, providing a winter abode for Newlyn's fishing fleet. Many Cornish ports have these historic harbour-works, small-scale on a map, but substantial up close. Usually too small to accommodate larger vessels, these little quays held smaller fishing boats and coastal craft, being man-made versions of (and in some cases extensions to) the natural havens which indent the Cornish coastline. Like the Fradgan cottages, they are profound expressions of the sheltering instinct. And none speaks of this better than Newlyn's old quay.

I ascend the Strand again as it turns and climbs steeply up the cliff; here there is a point to better overlook the old quay. Grass grows from the joints in the stones, while small boats rock contentedly in the harbour bowl. A small building of uncertain date stands where the quay joins the land: perhaps an ancient fishermen's chapel, since relegated to shed status? But what most impresses is the quay's almost hesitant curve as it reaches out into the sea, following no modern logic, obeying no formal geometries, just mounting its stonework against nature. But now, just as the Fradgan is enclosed by a more modern townscape, so the old quay is itself embraced by larger, later harbour works.

Newlyn emerges in the historical record in 1278 but is likely to be much older. It was once three separate seaside hamlets – Tolcarne, Street-an-Nowan and Newlyn Town – which gradually fused together in a sprawl of cottages, fish-cellars and little quays angling the shoreline. In the crook of the old quay could gather forty boats, perhaps fishing for as many households.

However, early Newlyn did not fish solely for subsistence. After John Cabot's discovery in 1497 of the vast shoals of cod in Newfoundland waters, boats from Newlyn sought profit there. Then in 1595, just as the port was growing, it was burnt to the ground by Spanish raiders, together with Penzance, Paul and Mousehole, as part of the post-Armada sabre-rattling between Spain and England.

Gradually they recovered, were rebuilt, and from the seventeenth century onwards Newlyn prospered again. With its roomy, sheltered anchorage on Gwavas Lake, it became a centre for the pilchard fishery, quasi-industrial in scale and fishing for surplus, not just subsistence. Millions of pilchards passed through its hulls and fish-cellars. These vast quantities were caught using ancient, little-changed methods.

Seining, for instance, was the skilful use of a bannerlike net, weighted at the bottom and floated at the top, to encircle a pilchard shoal and trap it by drawing the seine together into a purse. It required watchfulness and lightning-fast teamwork. Usually, a boat bearing the main net, accompanied by a few others, would lie inshore awaiting the pilchard shoal. The fishermen would not only be watching the water, however, but also gazing up at their colleague on the clifftop, who stood there scanning the sea's surface, waiting for the tell-tale brown, shimmering mass to flit towards the boats. When spotted, this 'huer' – as the spotters were called – would cry 'Hevva! Hevva!' and direct the boats with a pair of gorse branches. If all went to plan, the fish would run into the submerged net, which the other boats would then swiftly draw about the shoal. Once landed, these huge catches would be packed and salted, pressed to extract their oil and exported to Italy, where they were a vital staple for the many meatless days of the Catholic calendar.

Alternatively, nets would be drifted further out to intercept the shoal before it ran inshore, a little like trawling but static, much to the detriment of the seiners. Once the pilchard season had ended, that of the mackerel began. Then the methods changed to lining instead of seining: more languid, less tense, less surreally directed by men on clifftops with branches and trumpets. Long-lining was as it sounded: a long line bristling with hundreds of baited hooks, dangled down for mackerel.

Shoals waxed and waned, of course. Some years they were few, or didn't come at all. When this happened, starvation beckoned. Fishermen and their families had to subsist on the gleanings from the shoreline. But generally, Newlyn's pilchard fishery boomed through the eighteenth and early nineteenth centuries. Fish-houses multiplied – fish cellars on the ground floor, living quarters above. The quays were refaced, enlarged a little. Finer houses than cottages were built. But overall the place's pre-industrial character persisted. These changes were small-scale and enhanced rather than obscured the medieval waterfront and townscapes. Even in the mid-nineteenth century Newlyn could still be described in one guidebook as 'a colony of fishermen, with narrow paved lanes, glistening with pilchard scales in the season . . . [it] may call to mind the semi-barbarous habitations of some foreign countries – such as Spain'.[1] It's as if Newlyn's pre-industrial townscape mirrored the character of its pre-industrial fishery: nuanced, organic, formed with and not against nature's grain. More indiscriminate trawling was widespread elsewhere, but not, until the twentieth century, out of Cornish ports like Newlyn.

But as the nineteenth century ended, fishing boats were becoming larger and increasingly steam-powered. Vessels from the north-east coasts were drawn to the booming fisheries Newlyn had hitherto considered its own (and vice versa, it must be said); and by this time, the railway directly linked Cornwall with inland consumers, broadening the appetite for Newlyn's catch. Newlyn's winsome shoreline looked increasingly ill-equipped to cope with all this new business. Until now, catches had been landed on the old quay or on the beaches. But this only suited a limited fleet of smaller vessels. As for Newlyn's waterfront – well, that was a hopelessly irregular thing, a zigzag of small quays, inlets, rocks and housebacks. High tides cut off parts of the port from one another. And getting the fish off the beaches and through the narrow ways of its town was singularly awkward. So, inevitably, there came a programme of harbour-building, road-building and general 'improvement' in a typically Victorian style. Forthright new piers grew from the waterfront, the South Pier in 1886 and the North Pier in 1894. Enclosed by these massive structures, the old quay suffered no storms for the first time in four hundred years.

Newlyn's picturesqueness seemed to hang in the balance, but luckily it was caught in paint. From the 1880s, artists began to flock to the port. Led by Walter Langley and Stanhope Forbes, they were enchanted by the quality of light, the unspoiled views and the ancient streets of the Fradgan which satisfied the modern appetite for 'authenticity'. Their works caused a sudden awakening to Newlyn's ancient charisma, and bemused Newlyn fishing families found themselves immortalised in oils, paid more by the artists to stand still and be painted than they would have been for that day's catch. The paintings hung in London galleries and caused a sensation with their luminous depictions of a simple life.

Trawlers crawl methodically over a tract of sea much as a tractor does a field. Both drag behind them clouds of gulls and the same sense of settled subsistence on the world. Of the two, trawlers are the more vulnerable, pitting themselves against deeper odds. More than any other kind of vessel, they have an intimacy with the sea. They linger on it, plumb it, worry it and it worries them in return. And there is mysteriousness about them. They are the only vessels which disappear empty and reappear full without touching at any port. This would seem magical if we hadn't been fishing since prehistory. Fishermen are, as Cornish historian A.K. Hamilton Jenkin put it: 'the first of all the long-shore types to take to the sea for a livelihood, the last to leave it'.[2] Perhaps this venerability is reflected in the patronage of saints. While seafarers as a body are assigned St Nicholas, a land-lubbing ecclesiastic from Myra, fishermen claim no less than St Peter, former fisherman and premier apostle, with the keys to the gates of heaven.

The Newlyn School of painters captured the port on the

cusp of modernity. But, abetted by those Victorian improvements, the port's industrialisation broke its enchantment and the painters began to seep from the town; by the turn of the century, they had mostly disappeared, drawn to less blemished ports like St Ives. In 1904, like the smoothing out of an awkward crease, the zigzag way along Newlyn's shore was straightened and broadened into the Strand. It joined the two new piers and linked them in turn to Penzance and beyond. Lorries began to arrive, as they still do, for the great catches Newlyn's harbour could now handle. The town's economy grew, supplemented now by quarrying in its hinterland; minerals were loaded onto stone-boats from the new piers.

Amidst all this, Newlyn's original fishing community jostled for room. A measure of how greatly the character of its fishery had changed can be seen at the other end of the Strand. Here, in 1911, the Fishermen's Mission established a handsome granite outpost. Fully named the Royal National Mission to Deep Sea Fishermen, it sought to provide relief for the beleaguered crews of the steam-trawlers which worked out of Newlyn far offshore. These hissing, clanking vessels trawled deeply and indiscriminately, working against the grain of the shoals, without the nuance and harmony with the sea that had characterised the inshore fishery before. It was no surprise that most of the artists fled this new quayscape of iron and steam.

And, a few years later, of blood. Opposite the Fishermen's Mission is Newlyn's War Memorial, a simple granite cross with the names inscribed in Polyphant stone. Seventy-four men of the port lost their lives in the First World War, ploughed into muddy battlefields or sunk in the metal citadels which now passed for warships. Behind the names are stories of indescribable poignancy. To pick just one at random, to speak for the rest: William Maddern, five feet five inches, dark brown hair, grey eyes, non-swimmer, no known grave. Resident Gwavas Quay, Newlyn.[3]

At Tolcarne, the easternmost part of Newlyn, the waterfront suffered less from the Victorian and Edwardian improvements. Here was Newlyn's ancient boatbuilding ground. Here, on the beach, grew skeletons of fresh oak, planked, decked, painted, dragged down to the water's edge, and launched as fishing vessels. Up until 1919, the yard of the Peake family built these wooden boats along traditional lines, shaped for the rough seas off Land's End and the mackerel-drifting and long-lining there; they were the last of a long lineage. And that first post-war year, with Newlyn's population decimated by war but her seas at least calm again, the last of these vessels were launched from Tolcarne. One of them was *Rosebud*.

I'd taken the wrong turn to John's boathouse at first. Off from the street, a boathouse with pale blue doors, he said. But I'd spied a pair of pale blue doors down a lane into the Fradgan and was there diverted. Now I was hurrying back down the Strand, with the old quay on my seaward side and a group of industrial sheds to my landward. One of them, Trelawney's Fish, has a saw-tooth roof which looks like a child's drawing of a wave line. Behind Trelawney's is a service road to a garage. Angled at its head was the boathouse, one pale blue door ajar, silhouetting a man dressed as if for the deck of a lugger.*

Which, in fact, is about right; John looks like the long-suffering lugger-owner that he is. Years ago, he bought *Ripple*, an unrestored St Ives lugger of 1897 that had recently sunk at her moorings. Painstakingly he restored her on the quayside at Newlyn, and in 2007 relaunched her with the help of the local rugby team, who dragged her ceremonially into the sea. Since then he has voyaged in her to France and up and down the Channel ports, gathering with like-minded souls who know the madness of restoring and keeping such a vessel. By the look of him, *Ripple* seems to worry John every day, but it's clear that he adores her.

John looks as though he's seen a lot of sea. He tells me about his training as a master mariner, then his long career as an overseas civil servant in Hong Kong. It sounds as though he was quite senior there, and I find it hard to square the

* A kind of traditional fishing vessel unique to Cornwall.

image of him in a suit with the man in sailing gear now stand-ing with me in the chilly interior of the boathouse. The wind blows hard under the doors. A tiny little heater exhales a thin stream of warm air, hopelessly outgunned. All around are cob-webbed sailing paraphernalia and offcuts from his restoration of *Ripple*. For reasons beyond his control, John hasn't put out to sea this year; *Ripple* hasn't moved from her moorings and is a little green about the gunnels.

John grew up in nearby Mousehole, son of one of the painters who were drawn by the light to west Cornwall. After Hong Kong he returned to Newlyn, where he lives not far away from the boathouse in which we now stand. He bought it, he says, as a place in which to found and develop a lugger-centre, a training-school for vessel restoration and sailcraft. He's inter-ested in prolonging the traditional knowledge and skills which still hang about Newlyn like the cobwebs in the boathouse, hanging by slender threads but not yet entirely extinguished by the passing of their exponents or the differing values of gen-erations grown more remote from the sea.

The thing about sailing boats like *Ripple*, he says, is that it's all about working with the hands, using your body, manipu-lating materials which were once alive and are in a sense still living, not inert. *Ripple* is a slog to handle. He describes having to haul the lugsail about the mast every time the wind changes direction. Boats like these are testing to sail and even more worrisome to restore, but that is why their owners fall so hard for them. John speaks of the physicality of sailing like a man who has spent considerable time behind a desk. His hopes for his lugger-centre are to encourage more people to shrug off their inertia, to go out into the real world and to reengage with the tactility of wood and the devilry of water.

But there is an undercurrent to John's boathouse. Its blockwork walls cannot be earlier than the post-war era, but it stands on an old plot, with a threshold that slopes downwards to the sea. To me it seems a rebuilt version of an earlier boat-house. And John thinks, with something approaching certain-ty, that it was the place where *Rosebud* was built.

There is a sudden barrage of rain-squall on the boat-house roof. We raise our voices over the drumming roar. John gestures over to an antique table with bulbous legs, covered by a sheet of white plastic; upon this lies a piece of *Rosebud*. It is crooked, in shape like a near-flattened chevron. Having become dehydrated in storage, surface particles frass away easily under my fingers, but her heartwood remains sound.

The grain comes naturally to a point; this must original-ly have been a kinked branch of the tree selected by *Rosebud*'s builder as suitable for use as one of her ribs. And John selected this piece as a prize when she lay as a hulk on Dynamite Quay. Home from Hong Kong on holiday, he went to Lelant Saltings and sawed off this part of her frame as a keepsake. I pose the question I have come all this way to ask: why? It takes him a while to think, then to answer. 'I don't know,' he says a few times, perhaps inwardly turning the question over. Then he looks at me and flashes a toothy smile. 'I don't know,' he says again. 'A whim. Because of what she stood for.'

When, in 1580, the *Golden Hind* reappeared in the Channel after three years away, it was a fishing boat that Drake's crew hailed for news. In Britain's maritime history such boats are always there in the background, witnesses to the comings and goings of great ships. Occasionally, though, they come into the foreground, in improvised roles different from those for which they were designed. Usually this is in times of crisis. In the First World War, for instance, the requisitioned Lowestoft smack *Inverlyon* sank a German U-boat with her lone, three-pounder gun. More famous are the flotilla of 'little ships' which helped to evacuate the British Expeditionary Force from Dunkirk in the early stages of the Second World War.

Between these wars, in October 1937, *Rosebud* cast off from Newlyn's South Pier on the morning tide. She carried only fishermen, but they had no intent to fish. She carried 1,093 signatures appended to this plea: 'We the undersigned inhabitants of Newlyn and district wish to protest respect-fully and strongly against the wholesale destruction of our village'.[4] Leaving Mount's Bay, *Rosebud* steered uncharacteristically eastwards up the English Channel, into the foreground.

Mariners are but a plank's thickness from eternity, as the old saw goes. Whether hulls are formed of Cornish oak, like *Rosebud*'s, or Clydebank steel, like *Lusitania*'s, they are fragile things to pit against oceans. Vastly different as these vessels were in almost every respect – *Rosebud* was one-thirteenth of *Lusitania's* length – the greatest gulf between them was perhaps demographic: one was made to bear passengers of the highest society, while the other carried crew of the humblest. While *Lusitania*'s passengers were shielded from the sea in an illusion of landed comfort, *Rosebud*'s crew comfortlessly fished in the sea's grip.

They did, though, have one thing in common: both vessels had architectural counterpoints on land. Just as *Lusitania* mirrored the appearance of a grand hotel, so *Rosebud* unconsciously echoed the character of the tiny cottages in which her crews lived. As Ruskin observes: 'the fishing-boat . . . in the architecture of the sea represents the cottage . . . [it] swims humbly in the midst of broad green fields and hills of ocean'.[5] Their granite walls were counterpoints to the fragile hulls in which they made their living. And in contrast with the formal architecture of liners and hotels, fishing vessels and fishermen's cottages sprang from the same ancestral vernacular. They grew from the particulars of a place, *belonged* to their specific ports and plots, would make no sense anywhere else. And they spoke not of life's entertainments but its essentials.

Despite the modernisation of the shoreline, despite the terrible toll inflicted by the First World War, Newlyn's fishing community had remained intact, the fishermen's cottages encircled by new developments but otherwise unmolested. But then, in 1935, besuited figures were seen picking their way through the Fradgan, the Narrows, the Navy Inn and other ancient parts of the town. They made themselves agreeable enough but had the power to command entry to the granite cottages. They were health inspectors, and they were there to judge whether Newlyn remained a fit place in which to live.

The war accelerated the slum clearances that had been initiated in Britain by the Victorians. The period was characterised by zeal for public health reform, aimed at bettering the lot of the poor and the downtrodden. Sometimes this meant reforms to workplaces, as with Froude's hull designs and Plimsoll's mark. Sometimes it meant topographical improvements, as with the new lighthouses and piers pitted against the sea. And sometimes people simply needed saving from themselves. Temperance societies, for instance, urged workers to avoid the evils of drink. And health inspectors told them that their homes were in fact hovels.

They meant well, of course. After the First World War it was felt that the returning servicemen and women should have decent homes in which to live. What had the war been for, anyway, if not for a better future? Across the country, places like the Fradgan began to be assessed with narrowed eyes. In some hard-to-define way, these cottages represented the crabbed old world from which the war had come. It's interesting that while many soldier-poets of the war invoked a vernacular ideal of the England for which they were fighting, these vernacular cottages seemed not to fit into the new, post-war world. Just as noteworthy, a sizeable proportion of Newlyn's residents actually welcomed the move to modern accommodation and the conveniences it offered, even to the extent of signing a counter-petition against the clearances. And more remarkably still, many of the holiday cottages that have since gutted Cornish ports are locally owned. Like shoals, local histories can split, veer, disobey expectations.

Many of the fishermen's cottages in Newlyn had been homes to generations of families. A few were genuine hovels but the majority were spotlessly kept. Apart from the ineradicable damp which still affects them today, they made for decent homes. However, most of them were granite shells only – one might say mere hulls. They kept out the weather, but with no running water, plumbing or modern amenities they were glaringly deficient in the eyes of the health inspectors. Bare hulls could never be homes, no matter how cleanly kept. Therefore, despite the fact that they were still very much functioning homes, the cottages of the Fradgan and other areas were condemned.

Subsequently, proposals were announced for the wholesale clearance of Newlyn's tight-knit townscape. New houses were to be provided further up the hill on the Gwavas Estate – houses with thin walls of blockwork instead of granite, and gas cookers instead of glowing ranges. They would, however, be much more than just hulls. They would be fully plumbed, would incorporate internal toilets, would be fitted out with all modern conveniences. No room for fish-cellars, net stores or sail-lofts, of course, but wasn't the fishery declining anyway? Increasingly, the Newlyn boats faced competition in the Western Approaches, not only from east-coast trawlers, but also from European fleets. The dominance of the trawl as the main method of fishing disrupted the shoals offshore, scuppering inshore techniques like seining. And, worse, escalating hostilities on the international stage led to the embargoing of pilchard exports to Italy, for so long a staple of Newlyn's economy. Consequently, many Newlyn sons chose not to follow in their fathers' wakes.

Demolition orders were drawn up by Penzance Borough Council, the authority which in 1934 had gained responsibility for Newlyn's governance following the abolition of the immemorial parish system. Incidentally, there had always been a potent rivalry between the two towns; Penzance, with its more metropolitan airs, looked askance at its saltier neighbour to the west. Once the orders were made public, shock gripped Newlyn, then protest. Emissaries went to the officials who were directing the plans, but they proved unbending. The Fradgan was to be fragmented.

It was against this backdrop that *Rosebud* picked her way up the English Channel in that autumn of 1937. She was bound for Westminster Pier, to deliver the townsfolk's petition to

Sir Kingsley Wood, the Minister for Health. After a week-long voyage she rounded Kent, entered the Thames estuary and motored sinuously towards London. She caused a stir. News, of course, had spread by this point. Many knew her errand. And there was something about her air of ad hoc pluck that stirred solidarity among the Thames bargemen who watched her pass; much more widely, *Rosebud* represented a cause everyone could understand. In the roar of crewman Billy Roberts, when they were moored at last under the shadow of Big Ben: 'The Cornish boys have come to fight for their homes!'

There is a photograph of *Rosebud*, too famous not to mention, chugging up the King's Reach, the Palace of Westminster looming gravely in the background. The fishermen on her decks have their fists raised in defiance. Few, if any, of Mount's Bay pilchard drivers had ever been seen in this watercourse. And as she lay at the wharf she looked like a fish out of water and therefore more like herself than ever before. The metropolitan setting threw her homely lines into stark relief.

Rosebud was built with low sides, the better to haul over the netfuls of mackerel to wriggle on her decks; this was her principal fishing style, but long-lining also suited her. Like her sisters, she had a little wheelhouse, mainmast and mizzenmast, and two Thornycroft petrol paraffin engines to drive her in calms and up rivers. Her hull shape had evolved a little over generations but still conformed to an ancient pattern. She measured fifty-seven feet from stem to stern, a little less at deck level, and was seventeen feet six inches in the beam. Her lines came to a flattened ellipse at the stern and at the bow to a hatchet-point; she owed her hull to Cornish forests, her lines to Cornish eyes, and had had no testing-tank but the sea itself.

Steeped in tar, creosote, bitumen, her planking yielded easily to the fire. First licking blue between the boards, then swelling orange around the pile, the flames consumed *Rosebud* as her last owners stood and watched her burn, eyes narrowed against the blaze, chainsaws lying idle on Dynamite Quay. It had taken a long time to dry out her hull, saturated as it was with the brackish water of the Saltings and the perennial Cornish rain. It had not always been their intention to burn her. For a long while they had held the notion of restoring her to her former glory so that she would venture out again into the Western Approaches, perhaps even all the way to Westminster Pier once more. But time and money had defeated them, like so many others who dream of resurrecting old boats, many of which had rotted away as hulks here in this estuary. Dynamite Quay is paved with good intentions.

Half a century before, it had been much harder to demolish the ancient cottages that were her spiritual counterpoints. Those blade-blunting granite walls had refused to yield easily. Perhaps the only thing more intractable than the stones was the mortar in which they were set. It was like demolishing caves. The difficulty suggested it was the wrong thing to do. Nor were the new homes on the Gwavas Estate found to be much good. As well as their intrinsic deficiencies, to reach them from the harbour required an exhausting climb up a steep hill. Fortunately, however, only a small proportion of the demolition orders had been enacted, nibbling away at the edges of the Fradgan and other neighbourhoods. These ancient quarters stood as if fearful, waiting for news of *Rosebud*.

Her mission had been an equivocal success. The minister had entertained the fishermen to tea and had heard what they had to say. He vowed to consider the matter carefully. In the end, as historian Michael Sagar-Fenton[6] puts it, Sir Kingsley Wood's response was a masterly *political* one. It involved saving from demolition a handful of homes and regrading the rest so that their owners would receive more compensation. But Newlyn still faced a substantial facelift. Unhappily, with the new Gwavas Estate completed, the evictions began. Municipal workers moved from cottage to cottage, clearing, padlocking, shuttering. For the first time in its deep history, the Fradgan lay as emptily as I saw it on that tempestuous October day.

In the direct sense, *Rosebud* had failed to prevent the clearances from proceeding, although a handful of properties were reprieved. Indirectly, however, she netted an enormous amount of publicity and sympathy for her cause. Henceforth, if local authorities wanted to rework the social fabric they governed, they had to do so under more scrutiny. And, crucially, *Rosebud*'s mission bought valuable time. Penzance Borough Council were much delayed in carrying out their plans, which were finally shelved by the outbreak of the Second World War.

And then, with bitter irony, the Fradgan was found to be habitable after all. Displaced French and Belgian mariners were housed in these ancient neighbourhoods as part of the war effort. The Council paid for the refurbishment of the cottages and the piping-in of services to those which lacked them – something it had insisted was impossible before. Ancient Newlyn largely survived, though one or two more of the 'slums' were cleared, including a small part of the Fradgan in 1955. The true damage had been done, however. The fishermen and their families had been scattered from the harbour. Some managed to remain, but the old, interwoven community which had so transfixed the painters of the Newlyn School had been unpicked.

To be a little fairer to Penzance Borough Council, it wasn't just Newlyn that had undergone a facelift. Parts of Penzance itself were condemned and demolished, among them a little district to the east of the town. The story there was the same. Penwith, Adelaide and Camberwell Streets were home to a thriving community, but were also found slumlike by a health inspector and demolished in wartime. This was a great shame not least because, like the old quayside of Newlyn, Camberwell Street produced a remarkable vessel which might have gained it the same sort of national publicity and sympathy. Alas, though, *Truelove* was too strange to be seaworthy.

Built in 1906, she was conceived by a visionary from Lowestoft who decided to build a boat in his cottage. Robert Austin Ellis had come to Penzance as a railway-builder, settling his family on Camberwell Street, a linear, cottage-crammed lane running between housebacks. A photograph of Mr Ellis survives, and I wish I could have met him. He has the face of a holy man. And who else but a seer would have the wherewithal to clear his back room and back yard and build, to his own design, a fishing smack out of floorboards? Many of his Lowestoft kin fished out of Mount's Bay – controversially so, as part of an influx of East Anglian fishermen who competed with the Newlyn boats for the rich shoals here. And Ellis was clearly drawn more to the sea than to the railways. His boat, *Truelove*, was built lying on her starboard side, it having proved impossible to prop her vertically within the confines of his cottage. Her bows took shape inside, taking on the proportions of the back room, while her midships and stern grew to the proportions of his yard.

On the day of her launch, various walls were demolished to allow her to pass through the tangle of narrow streets to the shoreline. She was borne aloft on the shoulders of the residents of Camberwell Street and launched into Mount's Bay. She floated! Ellis sailed her once across Mount's Bay, but that was all. She was pronounced unseaworthy by the Board of Trade and beached as a curio on the edge of the town. I wonder how she would have fared at Westminster.

Strong in both *Truelove* and *Rosebud* is the synergy between a fisherman's home and his hull. They translate the same need into different materials. *Truelove* was drawn from a cottage in several senses. Her strange story would seem fictional were it not for the surviving photographs. Finished, she stuck out of Ellis's cottage as though fused with it, carrying the continuity between hull and home almost to literal lengths; and, in the end, she was too cottage-shaped to be true. *Rosebud*, on the other hand, did not have her lines actually dictated by a cottage, but stemmed from a similar vernacular. And her voyage to Westminster had been ingenious. For, out of her context, she became something more than a fishing boat; she became a representation of the shelter that her fishermen stood to lose. All of them lived in the condemned areas. Not only did she deliver a petition, she also delivered herself, the metonym of their faraway cottages in Newlyn. In a way she still lives on in them. I think back to the Fradgan, and the way its straggling

19. *Truelove* growing startlingly
from Mr Ellis's cottage

cottages seemed like boats at a harbour, frozen mid-jostle and
translated into stone.

Ice and entrails glisten in the soakaways. I feel like a trespass-
er here, under the glares of blade-wielding fishermen clad in
yellow rubber. And it's true that I'm a lubber who would flinch
from the visceral work they do, trawling and gutting, hauling
and cutting, mooring and spluttering out with exhaustion.*
Fumes from the trawlers and the lorries flicker in and out
of my nostrils, beaten back by sea air. Piled on the quay I see
crates of the silvery fish won like minerals from that perilous
realm beyond the harbour arms. Come to Newlyn, if only once,
to appreciate this.

  Newlyn is fresh, Mousehole frozen. Quaintly romantic
as I am, perhaps I really belong in the latter, picturesquely os-
sified as it is. I haven't the stomach for the business of fishing,

---

* Admirably chronicled by Lamorna Ash in her recent *Dark, Salt, Clear:
Life in a Cornish Fishing Town* (2020).

but I do welcome the catch when it's cleaned, cooked and art-fully presented, its provenance carefully noted. These days we do like our food to be *from* somewhere. It seems more mean-ingful that way, but in the case of fish I've always found this concern amusing. They are not from anywhere but unmapped fathoms, traceable only by the name of the landing port: this is another way, perhaps, in which we impose ourselves upon the sea's ways.

If you walk out of Newlyn towards Mousehole, the road narrows as you come upon Lower Green Street. This was anoth-er of the neighbourhoods threatened in the clearances. Parts of it fell, but a few cottages linger on. Double yellow lines painted on this narrow way draw your eye out into the distance, where more of the grey sea fills the view. At the end of the road, to sea-ward, are two unprepossessing bungalows, low enough for the horizon to hover over their rooflines in the background. They stand off from the road behind a low wall, cement-rendered, bearing a piece of *Rosebud*'s hull. It's mounted there unassum-ingly next to a little plaque, unseen for the most part by those who walk the streets she helped save. It is perhaps the offi-cial relic of her on public display. Underhand the wood is scaly from the salt and the rain, as if it has long gone untouched.

Rosebud Court, as these bungalows are known, were com-pleted in 2000 by Penwith Housing Association for the rental of local families. Since the 1930s the site had stood empty, cleared before *Rosebud* moored at Westminster. The eventual restoration of this plot for local people, pleasing though it is, was exceptional. After the war had ended, most of the formerly condemned cottages were sold on the open market – a trend gloomily echoed in the stories of other Cornish ports, with one almost inevitable outcome: holiday cottages.

To John, his birthplace of Mousehole is now barely alive, hollowed out as it has been by vacations in its ancient homes; Newlyn is now almost no better, in his opinion, with a year-round population but with increasing numbers of holiday lets appearing in the Fradgan and other areas. And the thing about them, he says, is that they're only set up for a week's worth of life. You can't do anything in them beyond a week's sight-seeing. John and I have an indignant conversation about this trend, which renders Cornish ports staler and staler until they expire as real places – if by real we mean self-sustaining. Yet perhaps it is more complicated than that. Like the tradition-al fishing boats which were their counterpoints, these cottages have come adrift in modern times. With many or all of them lacking gardens, parking spaces, and the sheer floorspace re-quired to store the stuff of modern life, it's hard to see how they could be altered to remain modern family homes without being razed and begun again. And perhaps the uncomfortable truth is that local people aren't just priced out of these places; they do not see their lives as being anchored to the harbour-side any longer.

Newlyn's prospering fishery is exceptional. Elsewhere, the true catch these days is no longer fish, but a creature prized and despised in equal measure. Where would Cornwall be with-out its holidaymakers? As the economy currently stands, in the doldrums. Some are hopeful about the mining of lithium in the county's abandoned workings, or resurrected fishing fleets freed from continental quotas. *Rosebud* voyaged to Westmin-ster on a strain of this same hope: hope that the old ways could be maintained, or at least accommodated with the new. Now, as then, the promise of continuity is equivocal. Yet I remain upbeat. Frozen as they might be, unpeopled and shipless, Corn-wall's ports still yield nourishment for the soul. And they rein-force Newlyn's uniqueness. Just as *Rosebud* edged clear of her peers, so Newlyn survives as one of a kind, as unlovely as all fishing ports should be.

At the war's end, in 1945, *Rosebud* was sold and renamed *Cynthia Yvonne* after the new owner's daughters. She continued to fish out of Newlyn, successfully, until sold again to fish from Hayle for crayfish. In the 1960s she was sold yet again, to a consortium with a scheme to raise the *Titanic*; surreally, this former long-liner was to hover miles above that sunken liner, acting as a kind of pilot-boat while divers conducted opera-tions below. Neither the scheme nor the liner floated. Then came a slow tumble through the hands of more owners, who aspired to her restoration but found her unquestionably past her prime. She had become to a seafarer what the Fradgan cottages were to those health inspectors: a poor hulk, stranded out of her time.

Before she was burned, a notice appeared in *The Cornishman* inviting her remembrancers to help themselves to souvenirs. Many did. And in my way I have taken something from her too, in the form of the stories told to me by John, and now from Mike too. After I have shivered myself blue in Newlyn harbour, Josa collects me and we drive up to Mike's shed, sunk deep in the green hinterland high above Mount's Bay. I leave her and Ida in the nearby pub and ring Mike's doorbell.

He's a former builder and maker of bottled ships (for no less than Harrods) who, now 'retired' (his ironic quote marks), writes books and restores old things. I like his warmth, his wryness, his wide interests, his idiosyncrasy. We go out to his shed, which is a trove of relics, mangled ironwork and vintage motorbikes: a perfect place to become absorbed in creative pottering. From a corner he heaves up a huge chunk of wood in both arms and lands it with a thump on his scarred workbench. It looks not unlike a railway sleeper with one end curved up.

This is *Rosebud*'s forefoot – the section of keel at the bottom of her prow, the first to cleave the water, the first to nudge the land. Mike thinks it's Cornish elm, now mellowed to a quite beautiful pale brown; though furrowed and flaked it is sound to the touch. He also produces a crate of her wrought-iron nails, which he scooped from the quayside and prised out of her strakes. Each would have been handmade by a blacksmith, and they are substantial things, most now kinked by their salvage. He gives me one as a token. It smells faintly of fire.

One Sunday, after a good walk along the beach at Porthkidney, he found himself at Dynamite Quay, stood over her with a handsaw. Shrewdly he assessed the unlit bonfire that *Rosebud* had become, seeing that some of her was no longer original – her decks, for instance, having been replaced rather crudely in modern times. So he decided to saw off a piece of her keel, certainly original and therefore a true relic. Like a house's foundation stone, the keel is never replaced. I ask him why he saved this piece of her. 'Why? If I don't, it will be gone forever, maybe nobody else, well – it's just quality, it's just *quality*, beautifully made, not fussily made, just made with skill, day-to-day, ingrained skill, where things just happened right in your hand . . .' No doubt he means that they happened correctly, but I also hear an echo of John's laments for life's lost tactility.

Mike drove *Rosebud*'s forefoot home from the Saltings, set it down outside his house and washed the salt from the wood. For a while it stayed outside, acclimatising, until he brought it into his shed. I ask him what, if any, were his ultimate intentions. 'I thought I'll do something with it, I'll make the legs for a bench, and every time I came to it with a saw I thought "I can't do that" . . . then I thought I'd run it through a saw and make the hulls for ships in bottles and then I thought, well I can't do that either, so I never actually did anything with it at all.'

What stayed his hand each time? Respect, he says, and the sense of it being too special to be reincarnated as another object. He tells me of old houses he's restored, his own in this village and an archaic place in Penzance –how he gutted these buildings, swept away their interiors and remade them. The ghosts were still in them, he says, and we should just have left them alone. I talk about how pleasing I find *Rosebud*'s material fate, ending up in fragments in people's sheds, splintered into hundreds of pieces cherished by individuals for one reason or another; fragmented, in fact, and subsumed back into the same community for which she once sailed. And I hasten to qualify myself: I'm just a non-seafarer, romanticising it all.

But Mike retorts: 'Well, I'm romanticising it because I'm keeping it! If I had any sense I'd chop it up and make it into firewood, to keep me warm for one night. I mean, my whole life is about romanticising the past, for goodness' sake . . .' We run our hands over *Rosebud*'s forefoot, feeling her enduring soundness underhand, the warmth in the wood: a warmth more enduring, in fact, than she might emit in the grate.

20. *Rosebud* arriving at
Westminster

Anchor

Amis Reunis

PORTMEIRION, WALES
TWENTIETH CENTURY

One last port now, where no ships put in, other than one destined never to leave.

I walk down a steep, wooded hillside sown with pastel buildings. As I round a curve in the road, groups of trees part to reveal a pretty quayside and, facing an estuary framed by mountains, a hotel with gables and bow windows. White parasols are planted along a terrace fringed with a white balustrade like the trimming on a garter. Poking up beyond it are a pair of masts.

From the terrace I drop down to the quayside, pausing to look back up at the baroque village perching on the hillside. Clouds pass quickly over the huge, estuarine sky like swimmers kicking anxiously over a deep abyss. I walk along the quay to where the ship lies jauntily at her mooring. She is painted in weatherworn white and her bowsprit points upriver.

I step from dock to deck. No change on the sole, no switch in motion, no buoyancy underfoot. What witchery is this, I wonder, then look up at the masts and see no canvas, or rope, but only wires slicing the sky into sail-shapes. I look down to the gap between hull and harbour wall, but there is none. There is only hardness and fastness: a boat grown from the stone. I stamp down hard on the stone-flagged deck, steadfast as the land, and feel the plot thicken.

Sailors have a curious expression for their retirement. To 'swallow the anchor', as they say, is to withdraw from sea life and settle down on land. I have not been able to track down an origin for this surreal maxim. Surreal, because an anchor is about the last thing you would think of attempting to swallow; think of those giant anchors to be seen gently rusting on quaysides throughout Britain, their sheer size, for one thing, but also their hard, sea-worn textures, their pitted paintwork, their dented shafts and blunted flukes (the barbed points at the curved ends).

When did Britain swallow the anchor? There are many possible moments, if moments alone had been enough to loosen the sea in our psyche. It may well have been the building of steamships which divorced us from the wind and the swell. Perhaps it was the new submarine warfare of the First World War, allegedly called 'underhanded, unfair and damned un-English' by an Admiral of the time;* or, indeed, the sinking of liners such as *Lusitania* where the sea's awful depths were exposed for the first time to vast crowds of civilians.

Or perhaps the fateful moment came during the Second World War, when the development of aerial warfare laid the groundwork for commercial aviation, lifting our eyes skywards. Or was it the rocketry of the 1950s and 1960s, which freed us to imagine and explore another fathomless realm beyond the earth? Or the commercial adoption of the jet engine in aeroplanes, which cheapened foreign holidays, leading our eyes towards the Mediterranean and away from our own coasts?

Forebodingly, on 31 July 1970, sailors in the Royal Navy received their last daily ration of rum, a date now lamented as 'Black Tot Day'; a few years later, Britain had joined the Common Market, which imposed quotas on the British fishing fleet, a source of continual strife ever since. Or perhaps the significant moment came in the next decade, when in 1982 HMS *Conqueror* sunk the *General Belgrano* during the Falklands War – the last time a British warship fired on an enemy vessel. Or perhaps it was all or none of these things. Perhaps these moments are largely meaningless. Perhaps a better question might be: *has* Britain swallowed the anchor?

*

* This quote, in wide and frequent use, is ascribed to Admiral Arthur Wilson, but its actual source is tantalisingly elusive.

'Soon we were down on the quay of the little enclosed harbour, a full-rigged ship alongside, lights reflected on the still water, the piled-up multi-coloured houses palely visible . . . indeed its image remained with me as an almost perfect example of the man-made adornment and use of an exquisite site – and a marine one at that . . .'[1] This architect was talking of Portofino, Italy, but his words are equally true of his own creation in North Wales: Portmeirion.

Planted on the northern side of the Dwyryd estuary, this curious village is the work of the architect Clough Williams-Ellis. Born in 1883, he was a child of the North Walian squirearchy, raised in a series of brooding houses on rugged mountainsides. Despite (or because of) this, he was a colourful, irrepressible figure, minded to be an architect from an early age. After a few false starts in the sciences, he enrolled at the Architectural Association, only to drop out when nepotism (cheerfully admitted) produced his first housebuilding commission. Further jobs came from the country-house weekends, balls and gala dinners at which he seemed perpetually to find himself.

Architecturally, he could work in many styles, often with great gravitas, always attuned to the special symbiosis between building and place. Above all, his buildings are *placeful*, playful, zesty, baroque: the last things, perhaps, that might be planted in this severe mountainside. Begun in 1925, opened a year later, added to for decades, Portmeirion is his masterpiece, a matchless lesson in placing buildings in a landscape, an illustration of how to build well in beautiful settings; a guide, in fact, to building anywhere without building obnoxiously. But Portmeirion is more than just an architecture lecture.

Portmeirion's idiosyncrasy is perhaps the result of its borrowings of forms and fabric from far beyond North Wales. Clough called it a 'home for fallen buildings': its Italianate domes, towers, gables and arcades incorporate bits of buildings salvaged from elsewhere. The result is, in the words of its creator, a place of 'wilful pleasantries . . . calculated naiveties, eye-traps, forced and faked perspectives, heretical constructions, unorthodox colour mixtures and general architectural levity'.[2] As landowner, architect, client and project manager, Clough had free rein to realise all this. Indeed, Portmeirion is perhaps so intoxicating because it speaks powerfully of this individual freedom.

Clough was not only an architect, but also a sailor. Describing himself as 'incurably sea-minded', as a boy he bought and renovated a ruinous dinghy. In 1920, he bought *Twinkler*, a small sloop in which he island-hopped, looking for Portmeirion's future site. Later, he acquired the *Scott*, a fifteen-ton Loch Fyne ketch which he 'bought rather cheaply because of her age and some dubious timbers'. In her, with his wife and three children, he sailed in and out of the Channel ports, skirted Ireland and made forays to the Highlands and Islands. On one particularly difficult night crossing from France, 'We three adults stayed on deck, too anxious to go below, taking turns at the kicking wheel and trying to keep the little ship more or less on course . . . at first dawn I clearly saw great breakers cascading back from black and jagged rocks . . .'[3]

Portmeirion was, in this sense, the final port of a long voyage brooding on the magic of coastal places. Though planted there by one man in his lifetime, the village somehow has the charisma of a place which has gradually developed, accreting the seaside by centuries instead of years. It shares this sense of happenstance with other coastal ports. It also shares, and takes to an extreme, their unreal lives as places dependent on tourism. Nobody, save perhaps Clough's descendants, could be said to be *from* Portmeirion; nobody lives there the year round. Time spent there is measured in hours, days, perhaps a week – measured in bright flashes of experience, perhaps, instead of the slow burn of residence.

Of course, Portmeirion does not have the drug problems, the joblessness, the dysfunctional housing market or the social tensions that wrack many coastal places in Britain today. It is not, like so many of them, a vacuum of lost prosperity. Rather, it exists in a kind of crisp and stilled perfection, like the magical ideal of a coastal town. In this it seems to distil our seafaring nostalgia. And it is an excellent place to ponder the question: what is our coastline now *for*?

The answer to this question, if you were to reckon the use of the sea by the types of boats now most commonly seen, is almost certainly *pleasure*. I think back to Plymouth, its ancient harbour now crammed with more fibreglass yachts than any other craft. Such harbour views would suggest that those of us who use the sea do so primarily as a pastime, rather than for any professional purpose. There are countless people like Clough who yacht incessantly in their spare time. I know many such 'incurably sea-minded' people, from the manager of my London apartment block to my Uncle Mark who, retiring from life in the fire brigade, recently bought a boat which he keeps at Torpoint.

In recent years, much has been made of Britain's 'sea-blindness' – the notion that because of commercial flight and freight carried in vast containerships we have lost our old sympathy with the sea. I am not so sure. Certainly, professional seafarers are not so commonly seen in Britain today. They are now more like a submerged, invisible demographic, concealed as they are by the impersonal traffic of bulk carriers and the container ports fenced off from where most live. But the countless leisure craft to be seen up and down our coastlines suggests that, when it comes to the sea, we are far from unseeing.

Those who use the sea for pleasure follow an illustrious wake. In 1651, hunted by Parliamentarians, Prince Charles was smuggled aboard an old collier named *Surprise* and ferried heart-stoppingly to France. *Surprise* was of thirty-four tonnes and had never carried anyone as famous as a prince. Her owner, Nicholas Tettersell, agreed to do it for sixty silver coins and a cover story.

After the Restoration in 1660, when that errant prince returned as King Charles II, he remembered *Surprise*. Purchasing and renaming her *Royal Escape*, he commissioned her into the navy and appointed Tettersell captain. In her, Charles sailed up and down the Thames and ventured out into the Channel. During the tribulations of the Civil Wars, Charles had learned to love the sea and gained experience of its ways; while in exile in Holland, he took to the Dutch custom of 'yacht-ing' – that is, sailing for its own sake, with no real view in mind other than that of the seascape. *Royal Escape* was one of the first yachts seen in Britain, and the only one of Charles's yachts not to have been purpose-built (the other twenty-five were scratch-built and sometimes experimental). Surprisingly, she remained a lowly part of the British fleet until 1791, by which time she had become so decrepit that she was broken up on the Thames foreshore.

Over the next few centuries, yachts were largely the pre-serve of royals, 'employed to convey princes of the blood, am-bassadors or other great personages from one part to another'. The largest, the sovereign's, was rigged with three masts like a ship; the others were rigged as ketches. Yet by 1769 there were also many smaller yachts, used by maritime officials or 'as pleasure-boats by private gentlemen'.[4] Now, there are many degrees of private boat-owning and sailing.

I saw this for myself at Scarborough Yacht Club. I went there to talk to the members about lighthouses, and I confess to feeling in awe of them – I am no sailor, yet I have dared to write on maritime themes. The yacht club occupies the light-house on the pier, affording fine views across the harbour and the sea beyond. From that sea, in 1914, came a bombardment from the German Grand Fleet, which rained shells onto Scar-borough town. This was no intended targeting of military sites, just a shelling of civilians to seed panic and probe the muscle of the Royal Navy. After my talk, the audience asked many sea-minded questions, which I parried as best I could, and after-wards they told me hospitably that I had done well. What had struck me about the club was how easy-going it seemed to be. I had thought yacht clubs to be quite aloof affairs, yet this one had an everyman feeling that I liked. The members talked of their seagoing in the unaffected terms of a pastime that gave them great joy. But the convivial atmosphere of the clubhouse stood in sharp contrast to the mood of the rest of the seafront, which was listless and downbeat. The fortunes of Scarborough as a seagoing town follow a familiar story. Only a handful of fishing vessels now contribute to the town's economy; few other vessels apart from yachts venture out beyond the harbour arms. Apart from this pastime, seagoing no longer seems a living activity.

And yet – I think of how the club's Commodore gleefully told me how they had recently re-enacted the bombardment of Scarborough with flares and formations of yachts keeping station offshore. And of how skippers such as John sink themselves deep into the upkeep and handling of historic ships like *Ripple*. And of how flotillas of yachts ride in anchorages all over Britain, collectively forming a fleet larger than that assembled by any formal navy. Often these are ancient anchorages, still in vibrant use. Downstream from the wreck of the *Grace Dieu* on the river Hamble, for instance, is a watercourse of thriving boatyards, marinas and the Royal Southern Yacht club. You'd be forgiven for thinking our seafaring prime lies not so far upstream.

I duck into the wheelhouse of *Amis Reunis*, as the stone boat is known. This had been the name of an old Porthmadog trading ketch which Clough Williams-Ellis had purchased in 1926, just after opening Portmeirion. It seems she was semi-permanently moored at the quayside as a kind of houseboat, with water and electricity laid on: 'she lay afloat in a fathom and a half or so . . . a most snug and romantic annexe'.[5]

One day, requiring new caulking and pitching, she was taken out to a sandbank in the estuary, stranded and careened. With the work done, she was moored a little way offshore in readiness to return to her permanent berth at the quayside. But then a sudden gale dragged and stranded her on a shelving shoal, lodged in such a way that when the tide ebbed she lay with her masts pointing almost horizontally, as though in mid-capsize. Clough did his best to save her with ingenious contrivances of barrels and winches, but she lay sand-fast, stubbornly beyond salvage. So instead he saved what timberwork and contents he could and ferried them to the new quay he was constructing for Portmeirion. There, under the hands of his masons, a curious effigy of her grew from the stone, bulging out from the quay and into the estuary, a ketch caught permanently by the river.

I walk over her details. Really, she is a sea-folly, distantly authentic, toy-like when boarded; a ship-simulacrum, permanently hove-to. Most of all I love her immovability, for she throws every other ship I have seen on my travels into sharp relief, makes them seem more *themselves*. She salts them with her own motionlessness, making them seem more movingly alive. In truth, I hadn't expected this journey to end here; I had expected to anchor in some once-thriving port, or at least somewhere more famously nautical, but there is not much evidence, in spite of the decoying presence of *Amis Reunis* that Portmeirion was ever much of a port. And yet, in 1946, this notice appeared in a guidebook:

> In view of the perpetual damage to boats and their *frequent actual loss* [my italics], Portmeirion guests are implored to exercise ordinary seaman-like care and at least see that they leave their craft securely moored with all gear safely stowed inboard.

Sometimes the history of seafaring can feel just out of reach, just around the corner, just a little way downstream. Vacant harbours like that of Mousehole, in their old pristineness, can feel as though their original fleets have just slipped them, poised to reappear over the horizon at any moment. Even far inland, the sea has a presence. Barely 100 metres from me, John Ruskin lived in a lost house in what is now a South London park. Although he wrote his *Harbours of England* while recuperating from illness in Tunbridge Wells, if he was anything like me he was brooding fiercely upon 'the blunt head of a common, bluff, undecked sea-boat'[6] while pacing the streets of Camberwell, where I have done the same thing. And as I walk Portmeirion's empty quay, the swarms of pleasure-craft knocking together feel simultaneously close – it feels so easy, in theory, to putter over from the opposite shore and moor up – and immeasurably far off. And although this echoing estuary now seems like the last place to find seafarers I could, it's true, imagine my *Asunder* limping upriver to dock at this whimsical port.

I talk to Rachel Hunt, Portmeirion's Collections Manager, about whether she has ever seen any seagoing vessels docking in her time at the village. No, she says, but for one curious episode. Years ago, with Clough long since passed away, she recalls seeing a small yellow boat named *Pluto* moor up at the quay. It was a fine day, possibly early autumn, with the estuary flooded with bright sunshine. It was very small, little bigger than a rowing boat, although she wonders whether her memory may have distorted the details. The crew of the *Pluto* seemed curiously confident. A pause. Pluto, she says, was Clough's nickname within the family.

Nostalgia is, by one definition, a sentimental longing for or imagining of a period of the past. And nostalgia for Britain's seafaring prime is marked from the twentieth century onwards in shipbuildings proudly advertising their timbers, buccaneering resorts in empty-harboured coastal ports, the glorification of fishermen in political wrangling. And, whether coincidentally or not, according to the *Oxford English Dictionary*, it is only from the twentieth century that nostalgia acquired this meaning. Originally, it described homesickness. During Captain Cook's first circumnavigation, the naturalist Sir Joseph Banks noted in his journal that by September 1770, off the Australian coasts and a year into the voyage: 'The greatest part of them [the ship's crew] were now pretty far gone with the longing for home which the Physicians have gone so far as to esteem a disease under the name of Nostalgia'.[7]

Sheer happenstance brought about the stone effigy of *Amis Reunis*. At the time of her loss, Clough noticed that the planned end of his new, half-built quay resembled the curve of her lost bulwarks, as though she had left something of herself on his drawing-board; so, he saved as much of her as possible and reconstructed her as 'more or less a ship-aground'.[8] By 1929 the new quay was finished, complete with a resurrected *Amis Reunis* merged into the sea wall. I step around to her stern, where a ship's wheel is placed below the mizzen. Idly I palm this helm, feeling it wobble loosely on its spindle. 'So there she is,' Clough later said, 'a mockery to seamen but a delight to adventurous children who can navigate her by twiddling her authentic steering wheel or agonize their parents by swarming up her dubious shrouds.'[9] Indeed, she reminds me of a ship-shaped apparatus in the playground near my home, over which myriad children, including my own, delightedly scramble. As such, the stone boat fulfils a function of which many yachters would doubtless approve: inspiring the sea in a younger generation.

But *Amis Reunis* seems to me to have a greater significance than this. For here, perhaps, is the answer to my guiding question: how should we remember the great ships of the past? Prolonging their lives unnaturally, as with HMS *Victory* and others like her, raises all sorts of philosophical questions – when every inch has been renewed, is she still the same vessel? – and many practical ones too. For ships are perishable, and it would have to be a nation sure of its priorities that undertook the pyrrhic battle with decay in every case. Yet simply remembering ships through paintings, literature and song doesn't seem quite enough, inadequate as these delightful things are to convey the ship's sheer physical quiddity. Is there a middle course to chart between the two?

Clough built the stone boat as a monument to the moment he swallowed the anchor. Only, when she was finished, that moment lay in the future. He seems to have voyaged throughout the 1930s until, with a growing family, the perils of seafaring seemed increasingly risky. While he said '[there is] much to be said for blood sports – when the blood is one's own',[10] his wife Amabel recalls being told that 'they would never make old bones'[11] if they continued. So, as the interwar years trailed to a close, Clough sold his *Scott* to a scrap dealer who used her to ferry spoil from abandoned coastal quarries.

At first, I think *Amis Reunis* memorialised his old ketch, but later began to represent the spectrum of vessels he had owned and skippered at one time or another, before finally becoming a symbol for the seafaring life that he had forsaken. By incorporating many of the original ship's fittings into her, he half-resurrected her with a degree of unquestionable authenticity, albeit in more durable materials; he made her form recognisably her own but with playful simplifications up close. Above all, she conveys her ancestor's size, shape and physical bearing in Clough's own whimsical manner; but, far more soberly, a shadow clings to her shrouds of what he called the 'blood sport' of seafaring.

As it seems to me, though Clough may not necessarily have erected the stone boat as an act of nostalgia, it nevertheless came to serve as a focus for it. And you might say that when it comes to our seafaring history, we as a nation are in a position approximate to the crew of *Endeavour* as observed by Joseph Banks all those years ago: yearning for a faraway ideal – the Britain of yesteryear, in the prime of its seafaring, flooded with the sea and its trappings. A Britain with the sea in all the senses: the sights of mast-forests, the smells of cargo heaped on quaysides, the tastes of exotica from the other hemisphere, the sound of seafarers thronging port towns, the tactility of wood or iron under the hand.

This Britain, of course, has dwindled away, displaced by a more nostalgic successor. And it's apt that yachts are now most numerous in our harbours and seaways. For the figurehead, as it were, of this pleasurable branch of seafaring is Charles II, a figure who embodies Britain's journey to naval supremacy but also its early slaving prime. He is a reminder of how a multitude of motives have always jostled at sea. In the past, our nostalgia fogged the most discomfiting, but now these are being raised to our view. We are beginning to look at our seafaring heritage with clearer eyes: now the challenge is to let no part of it, whether good or bad, ever sink again.

Famously, nowhere in Britain is very far from the sea, yet there is a curious shortening, lengthening, twanging of this distance now, as the sea has undoubtedly retreated from the lives of some, but has encroached far as a passion in the lives of others. Not much survives of the great ships of the past, but the instincts that sailed them are alive and well in marinas up and down the coasts. I think this everyday, commonplace kind of seafaring has more latency than we might realise. Britain now adjusts to a new geopolitical role in the world – a world that urgently needs low-carbon forms of transportation. Flight, terribly polluting in any case, looks even less appealing in a post-pandemic world. Much of this suggests that the sea, and the coastal places which have for so long lived strange half-lives, may be about to swing back into focus, that there may be a new calling for the seafarer's art.

Sailing is experiencing something of a resurgence. Cruise and container ships burn colossal quantities of oil and also contain terrific amounts of carbon in their structures. More pertinently, they encapsulate the overheated habits of a post-industrial Western world: cheap travel, cheap cargoes, cheap thrills – and as much of all of them as possible. Sailing may not only help to relieve stresses on nature, it may also help us to relieve the stresses on ourselves. In this, *Avontuur* is a particular name with which to conjure. This gaff-rigged schooner was built in 1920 in the Netherlands and, since 2016, has been running cargoes of coffee, rum and cacao across the Atlantic. Yes, it takes ages. But it is a taste of a bygone equilibrium with nature. And that somehow feels right.

It's perhaps fitting, then, that the anchor is also a symbol of hope – the hope that swells when it is lifted at the beginning of a voyage. I circle *Amis Reunis* one last time, looking for her anchor. Delightfully, she lacks one. She does not need one, of course: she *is* an anchor, a symbol of hope and, above all, of happenstance survival, fluke-like resurrection. Like all ships, she holds a sense of promise, and her example is one that we might consider adopting more widely around these shores. 'Part of the idea', Clough said, 'was that the sight of a considerable sea-going ship tied up to the wharf might suggest to seafarers that the little port was in business once again'.[12]

And I look at *Amis Reunis* there, lying up at the quay as though at any moment she might shake herself loose and innocently bear away for the sea.

21. *Amis Reunis* **berthed forever at Portmeirion**

Notes

Introduction

1. Samuel Rudder, *A History of Gloucestershire* (Cirencester, 1779), p. 524.
2. Ibid.
3. Quoted in F. H. Harris, 'Lydney Ships', *Transactions of the Bristol and Gloucestershire Archaeological Society for 1945*, 66, p. 238.
4. Quoted in David Cordingly, 'Willem [William] van de Velde, the younger', *Oxford Dictionary of National Biography* (Oxford University Press, 2004).
5. 'Letters and Papers Relating to the Navy, &c.: January 1658', in *Calendar of State Papers Domestic: Interregnum, 1657-8*, edited by Mary Anne Everett Green (London: HMSO, 1884), pp. 501-17.
6. 20 January 1673: 'Charles II: January 1673', in *Calendar of State Papers Domestic: Charles II, 1672-3*, edited by F. H. Blackburne Daniell (London: HMSO, 1901), pp. 380-503.
7. J. M. W. Turner, T. G. Lupton and John Ruskin, *The Harbours of England* (London: E. Gambart & Co., 1856).
8. Ibid.

Prow

1. Virgil, *Georgics*, trans. H. R. Fairclough, Book I, ll.136-8 (Cambridge, MA: Harvard University Press, 1916).
2. Quoted in George Thomas Clark, *Mediaeval Military Architecture in England*, Vol. II (London: Wyman & Sons, 1884), p. 12.
3. Caesar, *The Conquest of Gaul*, translated by S. A. Handforth (London: Penguin, 1951), p. 136.
4. According to the Roman poet Juvenal; see Juvenal, *The Sixteen Satires*, 3rd revised edn, edited and translated by Peter Green (London: Penguin, 1998), Satire IV, pp. xxiv, 28.
5. See The Migration Observatory report, October 2020; <https://migrationobservatory.ox.ac.uk/resources/commentaries/migrants-crossing-the-english-channel-in-small-boats-what-do-we-know/ >.

Trumpet

1. P. Marsden, *Ships and Shipwrecks* (London: Batsford/English Heritage, 1997), p. 69.
2. For a useful primer on this fascinating topic, see Eljas Oksanen, *Inland Navigation in England and Wales before 1348* (York: Archaeology Data Service, 2019); <https://doi.org/10.5284/1057497>.
3. William FitzStephen, 'A Description of London', in Henry Thomas Riley (ed.), *Liber Custumarum*, Rolls Series, No.12, Vol. 2 (London, 1860), pp. 2-15.
4. Jean Froissart, *Chronicles*, edited and translated by Geoffrey Brereton (London: Penguin, 1978), Book I.
5. Nicholas Antram, *Sussex: East with Brighton and Hove*, Pevsner Architectural Guides: Buildings of England (London: Yale University Press, 2013), p. 671.

6. Orders of Sir William Monson, quoted in N. A. M. Rodger, *The Safeguard of the Sea: A Naval History of Britain 660-1649* (London: HarperCollins, 1997), p. 320.
7. Ovid, *Metamorphoses*, translated by David Raeburn (London: Penguin, 2004), p. 22.

Trophy

1. *The World Encompassed by Sir Francis Drake*, compiled from notes by Francis Fletcher and others (London: Nicholas Bourne, 1628).
2. Ibid.
3. Martin Cortés, *The Arte of Nauigation Conteyning a Compendious Description of the Sphere, With the Making of Certayne Instruments and Rules for Nauigations, and Exemplifyed by Many Demonstrations*, translated by Richard Eden (London: Richard Watkins, 1589), Book 3, Chapter VII.
4. Ibid.
5. Ibid., Preface.
6. Michael Lok, 'Mr Lockes Discoors touching the Ewre, 1577' in Richard Collinson, *The Three Voyages of Martin Frobisher: In Search of a Passage to Cathaia and India by the North-West, A.D. 1576-8* (London: Hakluyt Society, 1867; reprinted by Cambridge University Press, 2010), pp. 92-3.
7. Collinson, *Three Voyages of Martin Frobisher*, p. 74.
8. 'Bisogna sapere adulare la natura' in Collinson, *Three Voyages of Martin Frobisher*, pp. 92-3.
9. Quoted in John Sugden, *Sir Francis Drake* (London: Pimlico, 2006), p. 98.
10. James McDermott, 'Edward Fenton', *Oxford Dictionary of National Biography* (Oxford University Press, 2004; updated 2008).
11. Hakluyt 3.679-79 in *Oxford Dictionary of National Biography* (Oxford University Press, 2006).
12. Sir Walter Raleigh, *The Last Fight of 'The Revenge' at Sea under the Command of Vice-Admiral Sir Richard Grenville, on the 10th-11th September, 1591*, edited by Edward Arber (London: A. Constable, 1895). See also Alfred, Lord Tennyson's poem, 'The Revenge: A Ballad of the Fleet'.
13. Nuno da Silva's sworn deposition in Zelia Nuttall (ed.), *New Light on Drake, A Collection of Documents relating to his Voyage of Circumnavigation, 1577-1580* (London: Hakluyt Society, 1914), p. 303.
14. 'Venice: June 1618, 16-30' in *Calendar of State Papers Relating to English Affairs in the Archives of Venice*, Vol. 15, 1617-1619, edited by Allen B. Hinds (London: HMSO, 1909), pp. 236-51; British History Online: <http://www.british-history.ac.uk/cal-state-papers/venice/vol15/pp236-251>; accessed 3 January 2021.
15. 'Drake's Drum' in Sir Henry Newbolt, *Poems: New and Old* (London: John Murray, 1912).
16. Matthew Baker, 'Fragments of Ancient English Shipwrightry', *c.*1570. The manuscript is now in the Pepys Library at Magdalene College, Cambridge.

Rope

1. Anon., *An Account of Several Workhouses for Employing and Maintaining the Poor* (London: Joseph Downing, 1725), p. 7.
2. N. A. M. Rodger, 'I Want to be an Admiral', *London Review of Books*, 42:15, 30 July 2020.
3. Sir Henry Mainwaring, 'The Seaman's Dictionary' (*c*.1622) in G. E. Manwaring and W. G. Perrin (eds), *The Life and Works of Sir Henry Mainwaring*, Vol. II (1922; reissued Abingdon: Routledge, 2019), p. 114.
4. 'Venice: June 1618, 16–30' in *Calendar of State Papers Relating to English Affairs in the Archives of Venice*, Vol. 15, 1617–1619, edited by Allen B. Hinds (London: HMSO, 1909), pp. 236–51; British History Online: <http://www.british-history.ac.uk/cal-state-papers/venice/vol15/pp236-251>; accessed 3 January 2021.
5. Samuel Pepys, *The Diary of Samuel Pepys, 1660–1669* (1825), entry for Monday, 3 August 1663; available at <https://www.pepysdiary.com/>.
6. Ibid., entry for Saturday, 22 June 1667.
7. N. A. M. Rodger, *The Safeguard of the Sea: A Naval History of Britain 660–1649* (London: HarperCollins, 1997), p. 384.
8. Ben Wilson, *Empire of the Deep: The Rise and Fall of the British Navy* (London: Weidenfeld & Nicolson, 2013), p. 242.
9. Edward Hasted, 'Parishes: Chatham', in *The History and Topographical Survey of the County of Kent*, Vol. 4 (Canterbury: W. Bristow, 1798), pp. 191–226; British History Online: <http://www.british-history.ac.uk/survey-kent/vol4/>.
10. J. M. W. Turner, T. G. Lupton and John Ruskin, *The Harbours of England* (London: E. Gambart & Co., 1856).
11. Capt. Frederick Chamier, *The Life of a Sailor* (London: Richard Bentley, 1850), p. 10.
12. Ibid.
13. Hasted, 'Parishes: Chatham', *History and Topographical Survey of the County of Kent*, Vol. 4, pp. 191–226.
    Daniel Defoe, *A Tour Through the Whole Island of Great Britain* (1724–6) (London: Penguin, 1971), 'Letter 2: Containing a Description of the Sea-Coasts of Kent, Sussex, Hampshire and of part of Surrey'; <https://visionofbritain.org.uk/travellers/Defoe/6> Quoted in Historic England website, entry for Chatham dockyard 'Sail loft' (1999); <https://historicengland.org.uk/listing/the-list/list-entry/1378586>.
14. See Matthew Symonds, 'Finding HMS *Namur*', *Current Archaeology*, 273, 2 November 2012.
15. Adapted from a description in Philip MacDougall, *Chatham Dockyard: The Rise and Fall of a Military Industrial Complex* (Stroud: The History Press, 2012).
16. Historic England website, entry for 'Former Rope Walk and Wall fronting Road at the Old Rectory', Aston Sandford, 11 October 1985; <https://historicengland.org.uk/listing/the-list/list-entry/1118342>.
17. MacDougall, *Chatham Dockyard*, Chapter 5.

18. This and the preceding quote are from William Falconer, *An Universal Dictionary of the Marine* (London: T. Cadell, 1769).
19. See various MSS in the National Archive and especially the Admiralty Papers, 106
20. Jonathan Coad, 'Historic Architecture of Chatham Dockyard, 1700-1850', *The Mariner's Mirror*, 68:2, 1982, pp. 133-185, see p. 141.
21. Ibid., p. 155.

Bell

1. Adapted from the original translation of Jacques Francis's testimony in the National Archives, High Court of Admiralty Papers: HCA 13/93, ff. 203-4.
2. Quoted in Miranda Kaufmann, *Black Tudors: The Untold Story* (London: Oneworld, 2017), Chapter 2.
3. Quoted in N. A. M. Rodger, *The Safeguard of the Sea: A Naval History of Britain 660-1649* (London: HarperCollins, 1997), p. 211.
4. James A. Williamson, *Hawkins of Plymouth: A New History of Sir John Hawkins and of the Other Members of his Family Prominent in Tudor England*, 2nd edn (Adam & Charles Black, 1969), p. 133.
5. Edmond Halley, 'The Art of Living Underwater . . .', *Philosophical Transactions*, 29: 349, 1714, p. 495.
6. E. S. Forster, *The Works of Aristotle translated into English . . .*, Vol. VII, *Problemata* (Oxford: Clarendon Press, 1927), Book XXXII, p. 5
7. Samuel Pepys, *The Diary of Samuel Pepys, 1660-1669* (1825), entry for Monday, 21 September 1668; available at <https://www.pepysdiary.com/>.
8. All quotes in this paragraph are from Thomas Phillips, *A Journal of a Voyage made in the* Hannibal *of London, Ann. 1693, 1694, from England, to Cape Monseradoe, in Africa . . .* (London: Henry Lintot and John Osborn, 1746), pp. 218-19.
9. Ibid., p. 236.
10. Ibid., p. 219.
11. This quote and the one preceding are from Phillips, *Journal of a Voyage made in the* Hannibal *of London*, p. 237.
12. Quoted in Halley's Log (anonymous blogger), 'Halley's maritime experience, part 1: Hally a Sayling', posted 22 June 2014; <https://halleyslog.wordpress.com/2014/06/22/halleys-maritime-experience-part-1/>.
13. Halley, 'The Art of Living Underwater . . .', pp. 497-9.
14. This and the following quotes are from Olaudah Equiano, *The Interesting Narrative of the Life of Olaudah Equiano, or Gustavus Vassa, the African* (London, 1789), Chapter 2.
15. Ibid., Chapter 3.
16. Quoted in James Oldham, 'William Murray, first earl of Mansfield', *Oxford Dictionary of National Biography* (Oxford University Press, 2004; updated 2008).
17. Quoted in Trevor Burnard, 'A New Look at the Zong Case of 1783', *XVII-XVIII*, 76, 31 December 2019; <http://journals.openedition.org/1718/1808>.

Notes

18. See the Centre for the Study of Legacies of British Slavery online database: <https://www.ucl.ac.uk/lbs/ >.
19. Madame d'Aulnoy, 'Le Prince Lutin' in *Fairy Tales by the Countess d'Aulnoy*, translated by J. R. Planché (London: G. Routledge & Co., 1855), p. 75.
20. Theories advanced by Tom Bennett and Peter C. Smith. See Peter C. Smith, *Sailors on the Rocks: Famous Royal Navy Shipwrecks* (Barnsley: Pen & Sword Maritime, 2015), Chapter 5, 'A Melancholy Echo'.
21. This quote and those following are from an account in *The Bristol Mercury*, 14 November 1896.
22. BBC News report, 'Edward Colston: Bristol Slave Trader Statue "Was an Affront"', 8 June 2020; <https://www.bbc.co.uk/news/uk-england-bristol-52962356>.

Figurehead

1. Homer, *The Odyssey*, translated by E. V. Rieu (London: Penguin, 1946), p. 191.
2. Revd John Troutbeck, *A Survey of the Ancient and Present State of the Scilly Islands* (Sherborne, Dorset: Goadby and Lerpiniere, 1794), p. 18.
3. William Borlase, *Observations on the Ancient and Present State of the Islands of Scilly, and their Importance to the Trade of Great-Britain* (Oxford: W. Jackson, 1756), p. 31.
4. Ibid. pp. 32–3.
5. As seen by John Evelyn and recorded in his diary entry for 9 April 1655 (*The Diary of John Evelyn*, 3 vols, edited by Austin Dobson (Cambridge University Press, 2015)).
6. Recounted in 'Ships' Figureheads' in *Shipping Wonders of the World* magazine, Part 25, 28 July 1936; <https://www.shippingwondersoftheworld.com/part25.html>.
7. Only relatively recently have they been systematically studied – see the work of Erica McCarthy ('Ships' Figureheads in Britain: An Evaluation of Their Changing Purpose and Interpretation Since 1750', unpublished PhD thesis, University of Hull, 2016), David Pulvertaft (*Figureheads of the Royal Navy* (Barnsley: Seaforth, 2011)) and the National Maritime Museum's Figureheads database.
8. J. A. Froude, 'On the Uses of a Landed Gentry: address delivered before the Philosophical Institution at Edinburgh, November 3, 1876', *Fraser's Magazine*, 14:84, December 1876, pp. 671–85 p. 281.
9. Letter to Sophia Tower, 28 November 1856, in Sophia Francis Tower (ed.), *In Memoriam. Scilly and its Emperor: Extracts of Letters from Augustus Smith, Esq., to S.F.T., 1845 to 1872* (Uxbridge: W. Morley, 1873).
10. W. Froude, 'On the Rolling of Ships', *Transactions of the Royal Institution of Naval Architects*, 2, March 1861, pp. 180–227, p. 181.
11. Samuel Plimsoll, *Our Seamen: An Appeal* (London: Virtue & Co., 1873), p. 72.
12. Ibid.
13. Revd I. W. North, *A Week in the Isles of Scilly* (London: Longman and Co., 1850), p. 122.
14. Letter to Sophia Tower, 1 January 1870, in Tower (ed.), *In Memoriam. Scilly and its Emperor: Extracts of Letters from Augustus Smith*.

Notes

15. Letter to Sophia Tower, 5 November 1870, in Tower (ed.), *In Memoriam. Scilly and its Emperor: Extracts of Letters from Augustus Smith*.
16. 'Serica', National Maritime Museum, Figureheads database entry; <http:// figureheads.ukmcs.org.uk/serica/>.

Timbers

1. This quote and those following are from the *Royal Cornwall Gazette*, edition of Friday, 19 November 1852.
2. Advertisement in the *Whitby Gazette*, 25 September 1856.
3. The quotes in this paragraph are from Ivor Stewart-Liberty, *Liberty's Tudor Shop, Great Marlborough Street* (London: Liberty & Co., 1949).

Mast

1. *Saunders's News-Letter*, Tuesday, 7 August 1787, quoted in 'John Wilkinson', *Grace's Guide to British Industrial History*; <https:// gracesguide.co.uk/John_Wilkinson >.
2. W. J. Lewis, *Ceaseless Vigil: My Lonely Years in the Lighthouse Service* (London: Chambers Harrap, 1970), p. 98.
3. Quoted in Bill Glover, 'Great Eastern', 2020; originally from L. T. C. Rolt, *Isambard Kingdom Brunel: A Biography* (London: Longmans, Green & Co., 1957).
4. Charles Dickens (ed.), 'A Morning Call on a Great Personage', *Household Words*, 2 January 1858, p. 62.
5. From a report by George Wilkes, editor of *Wilkes' Spirit of the Times*, quoted in Bill Glover, 2020
6. Rudyard Kipling, 'The Deep-Sea Cables' (1896).
7. Quoted in Anita McConnell, 'Sir John Pender', *Oxford Dictionary of National Biography* (Oxford University Press, 2004).
8. W. H. Russell, *The Atlantic Telegraph* (London: Day & Son,1865).
9. Sir Nathaniel Wraxall, *The Historical and the Posthumous Memoirs of Sir Nathaniel William Wraxall 1772–1784*, Vol. 2, edited by Henry B. Wheatley (London: Bickers & Son, 1884), p. 138.
10. 'The *Great Eastern* Steamship', *The Engineer*, 7 May 1886.
11. According to a report in the *New York Times*, 3 September 1869.

Propeller

1. Johnny Morris, 'The Night the Moon Came Down to Bathe', TV series *Tales of the Riverbank*, 1960.
2. Letter to Matthew Boulton, 1770.
3. Letter to the *Dublin Monitor*, 4 July 1840.
4. Letter from Augustus Smith to the editor of *The Mechanics' Magazine*, April 1854.
5. Report by *The Graphic*, 3 July 1897.
6. G. King and P. Wilson, *Lusitania: Triumph, Tragedy and the End of the Edwardian Age* (New York: St Martin's Press, 2015), p. 36.
7. *Dover Telegraph and Cinque Ports General Advertiser*, Saturday, 18 August 1838.
8. *Belfast News-Letter*, Thursday, 29 March 1888.

9. Quoted in *North British Daily Mail*, Thursday, 3 April 1890.
10. Cunard Board Minutes for 1908 (D42/B1/8).
11. As advertised in the *Western Daily Press*, Friday, 22 January 1937.
12. Eduardo de Martino, *The Naval Review at Spithead, 26 June 1897*, 1898, painting held in the Royal Collection.
13. Photographs from *The Hudson-Fulton Celebration, New York, 1909* (New York: Stereo-Travel Co., 1909).

Hull

1. T. C. Paris, *A Hand-Book for Travellers in Devon & Cornwall*, 4th edn (John Murray, 1859), p. 189.
2. A. K. Hamilton Jenkin, *Cornish Seafarers: The Smuggling, Wrecking and Fishing Life of Cornwall* (London: J. M. Dent & Sons, 1932), p. 137.
3. See the Roll of Honour website: <http://www.roll-of-honour.com/Cornwall/Newlyn.html>
4. Quoted in Michael Sagar-Fenton, *The Rosebud and the Newlyn Clearances* (St Agnes, Cornwall: Truran Books, 2003), p. 65.
5. J. M. W. Turner, T. G. Lupton and John Ruskin, *The Harbours of England* (London: E. Gambart, 1856).
6. His *The Rosebud and the Newlyn Clearances* is the essential account.

Anchor

1. Clough Williams-Ellis, *Architect Errant* (London: Constable, 1971), p. 193.
2. Ibid., p. 210.
3. Quotes in this paragraph, ibid., p. 22 and pp.194–6.
4. Quotes in this paragraph are from William Falconer, *An Universal Dictionary of the Marine* (London: T. Cadell, 1769).
5. Clough Williams-Ellis, *Portmeirion. The Place and its Meaning* (London: Faber & Faber, 1963), p. 58.
6. J. M. W. Turner, T. G. Lupton and John Ruskin, *The Harbours of England* (London: E. Gambart, 1856).
7. Sir Joseph Banks, *The Endeavour Journal of Sir Joseph Banks*, 3 September 1770; <https://gutenberg.net.au/ebooks05/0501141h.html>; II. 145 (1962). See entry for 'nostalgia' in the *Oxford English Dictionary*.
8. Williams-Ellis, *Portmeirion*, p. 59.
9. Ibid.
10. Williams-Ellis, *Architect Errant*, p. 201.
11. Christine Wallace Correspondence, March 2021.
12. Williams-Ellis, *Portmeirion*, p. 58.

Sources and Bibliography

General

Primary

British Newspaper Archive (BNA).

Calendar of State Papers: various monarchs and Interregnum, British History Online https://www.british-history.ac.uk/catalogue

Grace's Guide To British Industrial History (GG). https://gracesguide.co.uk/Main_Page

Samuel Johnson, *Dictionary* (1755). https://johnsonsdictionaryonline.org/index.php

National Historic Ships UK: The Registers (various vessels).https://www.nationalhistoricships.org.uk/

Samuel Pepys, *Diary* (1660-1669), reproduced at https://www.pepys-diary.com/

The National Archives (TNA): various MSS especially Admiralty papers

Secondary

Buildings of England; various counties and authors

J. J. Colledge, et al., *Ships of the Royal Navy: The Complete Record of all Fighting Ships from the 15th Century to the Present* (Seaforth Publishing, 2021).

Mark Dunkley, *Ships and Boats: Prehistory to 1840* (Historic England, 2016).

Mark Dunkley, *Ships and Boats: 1840 to 1950* (Historic England, 2016).

Historic England: various entries on the National Heritage List for England

Oxford Dictionary of National Biography; numerous entries (ODNB).

N. A. M. Rodger, *The Safeguard of the Sea: A Naval History of Britain 660-1649* (W.W. Norton, 1998).

N. A. M. Rodger, *The Command of the Ocean: A Naval History of Britain 1649-1815* (W.W. Norton, 2005).

Ben Wilson, *Empire of the Deep: The Rise and Fall of the British Navy* (Weidenfeld & Nicolson, 2013).

Rif Winfield, *British Warships in the Age of Sail 1603-1714* (Seaforth, 2009).

Rif Winfield, *British Warships in the Age of Sail 1714-1792* (Seaforth, 2007).

Rif Winfield, *British Warships in the Age of Sail 1793-1817* (Seaforth, 2010).

Sources and Bibliography

Introduction
Primary

'Charles II: January 1673', in *Calendar of State Papers Domestic: Charles II, 1672-3*, edited by F. H. Blackburne Daniell (London: HMSO, 1901).

'Letters and Papers Relating to the Navy, &c.: January 1658', in *Calendar of State Papers Domestic: Interregnum, 1657-8*, edited by Mary Anne Everett Green (London: HMSO, 1884).

Samuel Rudder, *A History of Gloucestershire* (Cirencester, 1779).

J. M. W. Turner, T. G. Lupton and John Ruskin, *The Harbours of England* (London: E. Gambart & Co., 1856).

TNA – various MSS in HCA 32/11.

Secondary

David Cordingly, 'Willem [William] van de Velde, the younger', *Oxford Dictionary of National Biography* (Oxford University Press, 2004).

F. H. Harris, 'Lydney Ships', *Transactions of the Bristol and Gloucestershire Archaeological Society for 1945*, 66.

Cyril Hart, *Industrial History of Dean* (David & Charles 1971).

Rif Winfield, *British Warships in the Age of Sail 1603-1714* (Pen & Sword Books 2010).

Prow
Primary

Caesar, *The Conquest of Gaul*, translated by S. A. Handforth (London: Penguin, 1951).

Juvenal, *The Sixteen Satires*, 3rd revised edn, edited and translated by Peter Green (London: Penguin, 1998).

Tacitus, *Agricola*

Virgil, *Georgics*, trans. H. R. Fairclough, Book I, ll.136-8 (Cambridge, MA: Harvard University Press, 1916).

Secondary

Kevin Booth, 'The Roman Pharos at Dover Castle' English Heritage Historical Review, 2:1, 8-21 (2007).

P. Clark, *The Dover Bronze Age Boat* (English Heritage, 2004).

Historic England *Ships and Boats: Prehistory to Present* (Historic England, 2017).

John Newman, *Kent: North East and East* (Pevsner Architectural Guides, Yale University Press, 2013).

Francis Pryor, *Britain BC: Life in Britain and Ireland before the Romans* (HarperCollins, 2003).

Sources and Bibliography

Trumpet

Primary

G. Chaucer and N. Coghill, *The Canterbury Tales* (Penguin Classics, 1951).

William FitzStephen, 'A Description of London', in Henry Thomas Riley (ed.), *Liber Custumarum*, Rolls Series, No.12, Vol. 2 (London, 1860).

Jean Froissart, *Chronicles*, edited and translated by Geoffrey Brereton (London: Penguin, 1978).

Ovid, *Metamorphoses*, translated by David Raeburn (London: Penguin, 2004).

Lotharingia law: Add. MS 14252 fols 99v-110r BL; http://www.bl.uk/manuscripts/FullDisplay.aspx?ref=Add_MS_14252

Secondary

Nicholas Antram, *Sussex: East with Brighton and Hove*, Pevsner Architectural Guides: Buildings of England (London: Yale University Press, 2013).

Thomas Asbridge, *Richard I: The Crusader King* (Penguin, 2018).

Simon Bradley, *London. 1: The City of London* (Penguin, 1997).

M. Champion, *Blackfriars Barn Undercroft, Winchelsea, East Sussex: Graffiti Survey Record* 2012

D. H. Farmer, (ed) *The Oxford Dictionary of Saints* (Oxford University Press, 1987).

I. Friel, *The Good Ship: Ships, Shipbuilding and Technology in England 1200-1520* (British Museum, 1995).

I. Friel, *Henry V's Navy: The Sea-Road to Agincourt and Conquest 1413-1422* (The History Press, 2015).

P. Marsden, *Ships of the Port of London: First to Eleventh Centuries AD* (English Heritage, 1994).

P. Marsden, *Ships of the Port of London: Twelfth to Seventeenth Centuries AD* (English Heritage, 1996).

P. Marsden, *Ships and Shipwrecks* (Batsford/English Heritage, 1997).

Jeremy Montagu, *The 14th century Billingsgate Trumpet* (Museum of London, 2017).

Eljas Okansen, *Inland Navigation in England and Wales before 1348: GIS Database* [data-set]. York: Archaeology Data Service [distributor] 2019 https://doi.org/10.5284/1057497 accessed 2020

N. A. M. Rodger, *The Safeguard of the Sea: A Naval History of Britain 660-1649* (W.W. Norton, 1998)

J. Schofield, L. Blackmore and J. Pearce with T. Dyson, *London's Waterfront 1100 to 1666: Excavations in Thames Street, London, 1974 -84* (Archaeopress, 2018).

Jonathan Sumption, *The Hundred Years War vol. 1: Trial By Battle* (Faber & Faber, 2011).

Jonathan Sumption, *The Hundred Years War vol. 2: Trial By Fire* (Faber & Faber, 2011).

John Webb, 'The Billingsgate Trumpet' in *The Galpin Society Journal* Vol. 41 (October 1988).

Sidney Painter, *The Third Crusade: Richard the Lionhearted and Philip Augustus*

Harry W. Hazard and Kenneth Meyer Setton (eds.) *A History of the Crusades vol. 2* (University of Pennsylvania Press, 1962).

Trophy

Primary

Matthew Baker, 'Fragments of Ancient English Shipwrightry', *c.*1570. The manuscript is now in the Pepys Library at Magdalene College, Cambridge.

Martin Cortés, *The Arte of Nauigation Conteyning a Compendious Description of the Sphere, With the Making of Certayne Instruments and Rules for Nauigations, and Exemplifyed by Many Demonstrations*, translated by Richard Eden (London: Richard Watkins, 1589).

Francis Drake, 'The World encompassed and analogous contemporary documents concerning Sir Francis Drake's circumnavigation of the world (1628; Editor: N. M. Penzer). With an appreciation of the achievement by Sir Richard Carnac Temple' (London: Argonaut Press, 1926).

Michael Lok, 'Mr Lockes Discoors touching the Ewre, 1577' in Richard Collinson, *The Three Voyages of Martin Frobisher: In Search of a Passage to Cathaia and India by the North-West, A.D. 1576-8* (London: Hakluyt Society, 1867; reprinted by Cambridge University Press, 2010).

Ben Johnson, et al. *Eastward Hoe* (1605) edited by Felix E. Schelling (D. C. Heath & Co, 1904).

Samuel Pepys, *Diary* 1660-1669. Reproduced at https://www.pepys-diary.com/

Sir Walter Raleigh, *The Last Fight of 'The Revenge' at Sea under the Command of Vice-Admiral Sir Richard Grenville, on the 10th-11th September, 1591*, edited by Edward Arber (London: A. Constable, 1895).

'Venice: June 1618, 16-30' in *Calendar of State Papers Relating to English Affairs in the Archives of Venice*, Vol. 15, 1617-1619, edited by Allen B. Hinds (London: HMSO, 1909), pp. 236-51. British History Online: <http://www.british-history.ac.uk/cal-state-papers/venice/vol15/pp236-251>; accessed 3 January 2021.

Sources and Bibliography

Secondary

R. C. D. Baldwin, *Borough, Stephen* (ODNB, 2008).

R. C. D. Baldwin, *Borough, William* (ODNB, 2008).

Margaret Blatcher, *Chatham Dockyard and a Little-Known Shipwright, Matthew Baker (1530-1613)* (Kent Archaeological Society, 2017).

Harry Kelsey, *Drake, Sir Francis* (ODNB, 2009).

Alexander Lindsay, *Cowley, Abraham* (ODNB, 2004).

David Loades, *Grenville, Sir Richard* (ODNB, 2008).

James McDermott, *Baker, Matthew* (ODNB, 2008).

James McDermott, *Fenton, Edward* (ODNB, 2008).

James McDermott, *Frobisher, Sir Martin* (ODNB, 2015).

Basil Morgan, *Hawkins, Sir John* (ODNB, 2007).

Sir Henry Newbolt, *Poems: New and Old* (John Murray, 1919).

Mark Nicholls and Penry Williams, *Raleigh, Sir Walter* (ODNB, 2015).

Nuttall, Zelia (ed) *New Light on Drake: a collection of documents relating to his voyage of circumnavigation, 1577-1580* (Hakluyt Society, 1914).

Anthony Payne, *Hakluyt, Richard* (ODNB, 2006).

Nikolaus Pevsner and Bridget Cherry, *Devon* (Penguin, ODNB 1989).

Rory Rapple, *Gilbert, Sir Humphrey* (ODNB, 2012).

Chris Robinson, *Plymouth's Historic Barbican* in E. Harwood and A. Powers, (eds.) *The Heroic Period of Conservation* (Twentieth Century Society, 2004).

N. A. M. Rodger, *The Safeguard of the Sea: A Naval History of Britain 660-1649* (W.W. Norton, 1998).

N. A. M. Rodger, 'Queen Elizabeth and the Myth of Sea-Power in English History'. Transactions of the Royal Historical Society, vol. 14, 2004, pp. 153-174. JSTOR, www.jstor.org/stable/3679312. Accessed 8 May 2021.

A. L. Rowse, *Sir Richard Grenville of the Revenge* (Jonathan Cape, 1937).

John Sugden, *Sir Francis Drake* (Pimlico, 2006).

Michael Turner, *In Drake's Wake* (Paul Mould, 2005).

Alden T. Vaughen, *American Indians in England* (ODNB, 2016).

Ben Wilson, *Empire of the Deep: The Rise and Fall of the British Navy* (Weidenfeld & Nicolson, 2013).

Rope

Primary

TNA - various MSS in ADM 106 relating to Chatham ropemakers.

Anon., *An Account of Several Workhouses for Employing and Maintaining the Poor* (London: Joseph Downing, 1725).

Sources and Bibliography

Capt. Frederick Chamier, *The Life of a Sailor* (London: Richard Bentley, 1850).

Daniel Defoe, *A Tour Through the Whole Island of Great Britain* (1724-6) (London: Penguin, 1971).

William Falconer, *An Universal Dictionary of the Marine* (London, 1769).

Edward Hasted, *History and Topographical Survey of the County of Kent* (1798).

Edward Hasted, 'Parishes: Chatham', *History and Topographical Survey of the County of Kent*, Vol. 4,. *British History Online* http://www.british-history.ac.uk/survey-kent/vol4/pp191-226

Sir Henry Mainwaring, 'The Seaman's Dictionary' (*c.*1622) in G. E. Manwaring and W. G. Perrin (eds), *The Life and Works of Sir Henry Mainwaring*, Vol. II (1922; reissued Abingdon: Routledge, 2019),

Samuel Pepys, *The Diary of Samuel Pepys, 1660-1669* (1825), entry for Monday, 3 August 1663; available at <https://www.pepysdiary.com/>.

J. M. W. Turner, T. G. Lupton and John Ruskin, *The Harbours of England* (London: E. Gambart & Co., 1856).

'Venice: June 1618, 16-30' in *Calendar of State Papers Relating to English Affairs in the Archives of Venice*, Vol. 15, 1617-1619, edited by Allen B. Hinds (London: HMSO, 1909); British History Online: <http://www.british-history.ac.uk/cal-state-papers/venice/vol15/pp236-251>; accessed 3 January 2021.

Secondary

Jonathan Coad, 'Historic Architecture of Chatham Dockyard, 1700-1850', *The Mariner's Mirror*, 68:2, 1982.

Historic England website, entry for 'Former Rope Walk and Wall fronting Road at the Old Rectory', Aston Sandford, 11 October 1985. <https://historicengland.org.uk/listing/the-list/list-entry/1118342>.

Philip MacDougall, *Chatham Dockyard: The Rise and Fall of a Military Industrial Complex* (Stroud: The History Press, 2012).

John Newman, *Kent: West and the Weald* (Pevsner Architectural Guides, Yale University Press, 2012).

N. A. M. Rodger, *The Safeguard of the Sea: A Naval History of Britain 660-1649* (W.W. Norton, 1998).

N. A. M. Rodger, *The Command of the Ocean: A Naval History of Britain 1649-1815* (W.W. Norton, 2005).

N. A. M. Rodger, *Nelson, Horatio* (ODNB, 2009).

N. A. M. Rodger, 'I want to be an Admiral' in *London Review of Books* Vol. 42 No. 15, 30 July 2020.

Ben Wilson, *Empire of the Deep: The Rise and Fall of the British Navy* (Weidenfeld & Nicolson, 2013).

Wessex Archaeology: *HMS Victory Conservation Management Plan*.

Bell

Primary

'Anthony Roll', British Library.

The Bristol Mercury, 14 November 1896.

D'Aulnoy, Madam Le Prince Lutin 1697.

Olaudah Equiano, *Interesting Narrative of the Life of Olaudah Equiano or Gustavus Vassa the African* (London, 1789).

Francis, Jacques, National Archives HCA 13/93, ff. 203-4.

Edmond Halley, *The Art of Living Underwater* in *Philosophical Transactions Vo. 29, No. 349* Royal Society, 1714.

Samuel Pepys, *Diary* 1660-1669. Reproduced at https://www.pepys-diary.com/

Thomas Phillips, *A Journal of a Voyage made in the Hannibal, 1693-5*

TNA:

PROB 11/1333/53. Will of Lancelot Skynner, November 1799.

MPH 1/812/2. Sketch of the Channel joining the Entrance into the Texel. The Soundings taken by the Master of the *Lutine* 21 August 99.

WO 32/8375. Lloyds offer of salvaged guns to City Corporation and Queen.

MT 10/556/6. Salvage operations on wreck of *Lutine*.

HCA 32/316/8/1-46. Captured ship: *Eendraght / Eendragt*, Pieter Gideon Udemans, master.

T 70/61. Instructions to Captains (2). Royal African Company ledger consisting of instructions to . . .

T 70/1433. The Black Book 1685-1702.

T 70/1223. Estimate of cargoes 1687-1703.

T 70/1222. Register of calculations . . . 1662-1699.

Secondary

W. R. Braithwaite, *Braithwaite, John, the elder* (ODNB, 2004).

W. R. Braithwaite and J. Bevan, *Siebe, (Christian) Augustus* (ODNB, 2007).

W. R. Braithwaite and J. Bevan, *Deane, Charles Anthony* (ODNB, 2011).

Trevor Burnar, *A New Look at the Zong Case of 1783 XVII-XVIII [Online], 76 2019*, posted on *December 31, 2019* http://journals.openedition.org/1718/1808

Centre for the Study of Legacies of British Slavery online database: https://www.ucl.ac.uk/lbs/ (accessed 2020).

Alan Cook, *Halley, Edmond* (ODNB, 2012).

Halley's Clerk (Anon Blogger) *Halley's Maritime Experience, Part 1: Hally a Sayling,* 2014. https://halleyslog.wordpress.com/2014/06/22/halleys-maritime-experience-part-1/

Historic England, List Entry for *Statue of Edward Colston* (accessed 2021).

Miranda Kaufmann, *Black Tudors: The Untold Story* (London: Oneworld, 2017).

Reyahn King, *Belle [married name Davinier], Dido Elizabeth* (ODNB, 2020).

Kenneth Morgan, *Colston, Edward* (ODNB, 2020).

James Oldham, *Insurance Litigation Involving the* Zong *and Other British Slave Ships, 1780-1897* (The Journal of Legal History 28:3, 299-318, 2007).

James Oldham, *Murray, William, First Earl of Mansfield* (ODNB, 2008).

Peter C. Smith, *Sailors on the Rocks: Famous Royal Navy Shipwrecks* (Pen & Sword Maritime, 2016).

John Sugden, *Sir Francis Drake*, (Pimlico, 2006).

James A. Williamson, *Hawkins of Plymouth: A New History of Sir John Hawkins and of the Other Members of his Family Prominent in Tudor England*, 2nd edn. (Adam & Charles Black, 1969).

Figurehead

Primary

Anon. *Ship's Figureheads* in *Shipping Wonders of the World* part 25, 1936

William Borlase, *Observations on the Ancient and Present State of the Islands of Scilly, and their importance to the trade of Great-Britain. In a letter to the Reverend Charles Lyttelton, LL.D., Dean of Exeter.* (Oxford: W. Jackson, 1756).

The Diary of John Evelyn, 3 vols., edited by Austin Dobson (Cambridge University Press, 2015).

J. A. Froude, 'On the Uses of a Landed Gentry: address delivered before the Philosophical Institution at Edinburgh, November 3, 1876', *Fraser's Magazine*, 14:84, December 1876.

W. Froude, 'On the Rolling of Ships', *Transactions of the Royal Institution of Naval Architects*, 2 March 1861.

Revd. I. W. North, *A Week in the Isles of Scilly* (Longman and Co., 1850).

Samuel Plimsoll, *Our Seamen: An Appeal* (London: Virtue & Co., 1873).

Augustus Smith, *Thirteen Years' Stewardship of the islands of Scilly, from 1834 to 1847* (London, 1848).

Letter to Sophia Tower, 5 November 1870, in Tower (ed.), *In Memoriam. Scilly and its Emperor: Extracts of Letters from Augustus Smith.*

Snorri Sturluson, *Heimskringla*, or *The Chronicle of the Kings of Norway*.

Revd. John Troutbeck, *A Survey of the Ancient and Present State of the Scilly Islands*, 1794.

Secondary

Bella Bathurst, *The Lighthouse Stevensons* (Harper Perennial, 2007).

P. Beacham and N. Pevsner, *Cornwall* (Yale University Press, 2014).

R. L. Bowley, *The Fortunate Islands: The Story of the Isles of Scilly* (Bowley Publications, 2004).

D. K. Brown and A. Lambert, *Froude, William* (ODNB, 2012).

Figureheads database, National Maritime Museum http://figure-heads.ukmcs.org.uk/

David Boyd Haycock, *Borlase, William* (ODNB, 2006).

E. Inglis-Jones, *Augustus Smith of Scilly* (Faber, 1969).

Peter Mandler, *Smith, Augustus John* (ODNB, 2004).

Erica McCarthy, *Ships' Figureheads in Britain: An Evaluation of their Changing Purpose and Interpretation since 1750 (*unpublished PhD thesis, University of Hull, 2016).

G. F. Matthews, *The Isles of Scilly: A Constitutional, Economic and Social Survey of the Development of an Island People from Early Times to 1900* (George Ronald, 1960).

Anita McConnell, *Plimsoll, Samuel* (ODNB, 2013).

Tom Nancollas, *Seashaken Houses: A Lighthouse History from Eddystone to Fastnet* (Particular Books, 2018).

Mike Nelhams, *Tresco Abbey Garden: A Personal and Pictorial History* (Truran, 2000).

A. F. Pollard and Rev. William Thomas *Froude, James Anthony* (ODNB, 2009).

David Pulvertaft, *Figureheads of the Royal Navy* (Seaforth, 2011).

Elisabeth Stanbrook, *Bishop Rock Lighthouse* (Twelveheads Press, 2008).

'William Froude' in GG, accessed 2020 https://www.gracesguide.co.uk/William_Froude

C. C. Vyvyan, *The Scilly Isles* (Robert Hale Ltd., 1953).

Timbers

Primary

Royal Cornwall Gazette, edition of Friday, 19 November 1852.

Stephen Gavin, interview conducted and transcribed Spring 2020.

The Evening News, 27 October 1926.

Advertisement in the *Whitby Gazette*, 25 September 1856.

Secondary

Susie Barson, *The Liberty Shops* (English Heritage, 1999).

P. Beacham and N. Pevsner, *Cornwall* (Yale University Press, 2014).

J. J. Colledge et al. *Ships of the Royal Navy: The Complete Record of all Fighting Ships from the 15th Century to the Present* (Seaforth Publishing, 2021).

Stephen Gavin, 'Lost Brig *Elizabeth Jane*' website http://lostbrig.net/
© Stephen Gavin

Historic England List Entry: *Church of St Nicholas* https://historic-england.org.uk/listing/the-list/list-entry/1201132

P. O. and A. V. Leggat, *The History of St Nicholas Church West Looe* (1997).

Neil Oliver, *A History of Scotland* (Weidenfeld & Nicolson, 2009).

Jeremy Pearson, *A Cornish Connoisseur and Builder: Lieutenant-Colonel Charles Lygon Cocks (1821–85)* in P. Holden, (ed.) *New Research on Cornish architecture* (Francis Boutle Publishers, 2017).

G. D. Rawle and G. Shaw, (rev) *Liberty, Sir Arthur Lasenby* (ODNB, 2014).

Ivor Stewart-Liberty, *Liberty's Tudor Shop, Great Marlborough Street* (Liberty & Co, 1958).

Ben Wilson, *Empire of the Deep: The Rise and Fall of the British Navy* (Weidenfeld & Nicolson, 2013).

Mast
Primary
'A Telegraphic Evening Party' in *The Illustrated London News*, 2 July 1870.

'The Great Eastern Steamship' in *The Engineer*, 7 May 1886.

Samuel Johnson, *Dictionary* 1755 https://johnsonsdictionaryonline.org/index.php

Rudyard Kipling, *The Deep-Sea Cables* (1896).

W. J. Lewis, *Ceaseless Vigil* (Harrap, 1970).

W. H. Russell, *The Atlantic Telegraph* (London, 1865).

'Sale of the Great Eastern' in *Liverpool Mercury*, November 21 & November 26 1888

Saunders's News-Letter, Tuesday, 7 August 1787, quoted in 'John Wilkinson', *Grace's Guide to British Industrial History*; <https://gracesguide.co.uk/John_Wilkinson >.

'The British Indian Submarine Telegraph' in *The Illustrated London News*, 20 November, 1869.

'The Great Eastern in the Mersey', in *Liverpool Mercury* Monday 27 August 1888.

'The Great Eastern: "All That Is Left Of Her"', in *Liverpool Mercury* 31 July 1890.

Jules Verne, *Around the World in Eighty Days* (1872).

Secondary
D. K. Brown, *Russell, John Scott* (ODNB, 2012).

R. A. Buchanan, *Brunel, Isambard Kingdom* (ODNB, 2017).

Bill Burns, (ed.) *History of the Atlantic Cable & Undersea Communications: From the first submarine cable of 1850 to the worldwide fiber optic network* accessed 2020 https://atlantic-cable.com/

Randolph Cock, *Schank, John* (ODNB, 2008).

J. Dugan, *The Great Iron Ship* (Hamish Hamilton, 1953).

Bill Glover, *Great Eastern* November 2020 in Bill Burns (ed.) *History of the Atlantic Cable & Undersea Communications: From the first submarine cable of 1850 to the worldwide fibre optic network* accessed 2020 https://atlantic-cable.com/

J. R. Harris, *Wilkinson, John* (ODNB, 2013).

Historic England List Entry *Site of the launch ways of the SS Great Eastern*, accessed 2020. https://historicengland.org.uk/listing/the-list/list-entry/1423608

'John Wilkinson' in GG https://www.gracesguide.co.uk/John_Wilkinson, accessed 2020.

Anita McConnell, *Pender, Sir John* (ODNB 2004).

'PS Aaron Manby' in GG https://gracesguide.co.uk/PS_Aaron_Manby accessed 2020

R. B. Prosser and Giles Hudson (rev) *Manby, Aaron* (ODNB, 2004).

L. T .C. Rolt, *Isambard Kingdom Brunel: A Biography* (Longmans, Green & Co., 1957).

'Southern Millwall: Drunken Dock and the Land of Promise', in *Survey of London: Volumes 43 and 44, Poplar, Blackwall and Isle of Dogs*, ed. Hermione Hobhouse (London, 1994), pp. 466–480. British History Online http://www.british-history.ac.uk/survey-london/vols43-4/pp466-480 [accessed 18 May 2021].

'SS Great Britain' in GG accessed 2020 https://gracesguide.co.uk/SS_Great_Britain

'SS Great Eastern' in GG accessed 2020 https://gracesguide.co.uk/SS_Great_Eastern

'SS Great Western' in GG accessed 2020 https://gracesguide.co.uk/SS_Great_Western

Ben Wilson, *Heyday: Britain and the Birth of the Modern World* (Weidenfeld & Nicolson, 2016).

Propeller
Primary

J. A. Bourne, *Treatise on the Screw Propeller, Screw Vessels and Screw Engines* (Longmans, Green and Co., 1867).

Cunard Archive MSS.

D42/S7 Files of secretary's correspondence and working papers on matters concerning the construction, outfitting, inspection, insurance etc. of the *Mauretania, Lusitania, Laconia, Aquitania, Andania* and *Alaunia*.

D42/B1/8 Board minutes for 1908.

'On Some Points of Interest in Connection with the Design, Building and Launching of the *Lusitania*' *The Engineer,* 12 April 1907.

'The Cunard Liner *Lusitania*' *The Engineer* 8 June 1906.

Secondary

G. C. Boase and W. Johnson, (rev) *Lowe, James* (ODNB)

David K. Brown, *Smith, Sir Francis Petit* (ODNB, 2013).

D. Finamore and G. Wood, *Ocean Liners: Speed and Style* (V&A Publishing, 2018).

'Francis Pettit Smith' in GG accessed 2020 https://www.gracesguide.co.uk/Francis_Pettit_Smith

C. Gibb and A. McConnell, *Parsons, Sir Charles Algernon* (ODNB, 2004).

Historic England List Entry: *The Boat House*, accessed 2020. https://historicengland.org.uk/listing/the-list/list-entry/1217493

G. King and P. Wilson, Lusitania*: Triumph, Tragedy and the End of the Edwardian Age* (New York: St Martin's Press, 2015).

'Leonard Peskett' in GG accessed 2020 https://www.gracesguide.co.uk/Leonard_Peskett

B. M. E. O'Mahoney and Roger T. Stearn (rev.) ,*Vansittart [nee Lowe], Henrietta* (ODNB, 2012).

'SS *Archimedes*' in GG accessed 2020 https://www.gracesguide.co.uk/SS_Archimedes

The *Lusitania* Resource (2011) https://www.rmslusitania.info/ accessed 2020.

Hull

Primary

T. C. Paris, *A Hand-Book for Travellers in Devon & Cornwall*, 4th edn (John Murray, 1859).

J. M. W. Turner, T. G. Lupton and John Ruskin, *The Harbours of England* (London: E. Gambart, 1856).

Secondary

Lamorna Ash, *Dark, Salt, Clear: Life in a Cornish Fishing Town* (Bloomsbury, 2020).

P. Beacham and N. Pevsner, *Cornwall* (Yale University Press, 2014).

'Cynthia Yvonne', 'Rosebud' and 'Truelove' in *PZ Fishing Boats List* Newlyn Archive, September 2018 hattps://www.newlynarchive.org.uk/pz-fishing-boats-list/ accessed 2020

Fiona Fleming, *Cornish Ports and Harbours Assessment: Newlyn Final Report* (Historic England, 2016).

Paul Greenwood *Once Aboard A Cornish Lugger* (Polperro Heritage Press, 2007).

A. K. Hamilton Jenkin, *Cornish Seafarers: The Smuggling, Wrecking and Fishing Life of Cornwall* (J. M. Dent & Sons, 1932).

'Ripple' *National Historic Ships UK* https://www.nationalhistoric-ships.org.uk/register/1960/ripple accessed 2020

Michael Sagar-Fenton, *The Rosebud and the Newlyn Clearances* (Truran, 2003).

Anchor

Primary

Rachel Hunt, correspondence March–May 2021.

Christine Wallace, correspondence March 2021.

Clough Williams-Ellis, *Portmeirion: The place and Its Meaning* (Faber & Faber, 1963).

Clough Williams-Ellis, *Architect Errant* (London: Constable, 1971).

Secondary

Lionel Esher, *Ellis, Sir (Bertram) Clough Williams* (ODNB 2014).

William Falconer, *An Universal Dictionary of the Marine* (London: T. Cadell, 1769).

Andrew Lambert, *Wilson, Sir Arthur Knyvet, third baronet* (ODNB, 2008).

'Nostalgia' in Oxford English Dictionary accessed 2021.

Ben Wilson, *Empire of the Deep: The Rise and Fall of the British Navy* (Weidenfeld & Nicolson, 2013).

Rif Winfield, *British Warships in the Age of Sail 1603–1714* (Seaforth, 2009).

Illustration Credits

1. A view of an unidentified British fifth-rate frigate. (National Maritime Museum, Greenwich).
2. The Dover Boat. (Historic England).
3. The medieval seal of Winchelsea. (Corporation of Winchelsea).
4. The Billingsgate Trumpet. (MOLA).
5. A drawing by Matthew Baker of *c.*1580 showing a 'race built' galleon. (By permission of the Pepys Library, Magdalene College Cambridge).
6. The Drake Chair. (Historic England).
7. *Victory* in drydock. (Author Photograph).
8. The Ropery interior. (Author Photograph).
9. *Jesus* from the Anthony Roll of the 1540s. (By permission of the Pepys Library, Magdalene College Cambridge).
10. The Lutine Bell, 1915. (Historic England).
11. *Rosa Tacchini* and the *Mary Hay*. (Author photograph).
12. *Valhalla*, turn-of-the-century photograph of the unrestored figureheads. (Courtesy of Tresco Estate).
13. The Ramscraigs croft house. (Crown Copyright: HES).
14. Interior of the church of St Nicholas, West Looe. (Baz Richardson).
15. *Great Eastern*. (Victoria and Albert Museum, London).
16. The lone mast from *Great Eastern*. (The Liverpool FC Collection).
17. First Class Dining Saloon, *Lusitania*. (Bedford Lemere & Co. Courtesy of DeGolyer Library, Southern Methodist University).
18. *Lusitania* propeller. (National Museums Liverpool).
19. *Truelove* growing startlingly from Mr Ellis's cottage. (By kind permission of Morrab Library).
20. *Rosebud* arriving at Westminster. (By kind permission of Morrab Library).
21. *Amis Reunis* berthed forever at Portmeirion. (Photograph courtesy of John S. Turner).

Acknowledgements

This book swam out of conversations with my matchless agent, Carrie Plitt, and inspiring former editor Cecilia Stein. To both – thank you indeed.

At Penguin, my current editor Chloe Currens wisely and peerlessly steered the book with me to completion. We had many fun and thought-provoking discussions along the way. My thanks also go to Thea Tuck and Anike Wildman for additional editorial help. Rebecca Lee, Steven Lovatt and Jane Robertson brought impeccable attention and guidance to the text, while Katy Banyard, Pete Pawsey, Richard Green and Jim Stoddart have created a wonderful book simply to look at. Thank you also to Auriol Griffiths-Jones for the index. I'd like to thank Matt Hutchinson, Liz Parsons, Thi Dinh, Ingrid Matts and Jon Parker for all their help in promoting the book, while the magnificent Chris Wormell has (as usual) given it a spectacular cover.

My thanks go to the following people who provided much invaluable insight and help at all stages of the book's gestation: Avril Kear, Jane Sidell, Roy Porter, Ian Friel, Jo Turner, John Spencer, John Schofield, Justin Reay, Catherine Sunderland, Will Nixon, the Master Ropemakers at Chatham, the Medway Archives, the Kent Archives, Serena Cant, Chris Pater, Graeme Sprowson, Amanda Martin, Mike Nelhams, Erica McCarthy, Stephen Gavin, Jeremy Pearson, Kathryn from Liberty, Bill Burns, Bill Glover, Simon Weedy, Keith Gagen, Sian Wilks, Mark Penrose, John Lambourn, Rachel Hunt, Richard Haslam, Edward Burness and Philip Chatfield.

For reading some or all of the book in varying forms, I owe a vast debt to Paul Robertshaw, Matthew Cooper, Nick Hayes, Ed Burness, Chris and Sue Nancollas and Margarette Lincoln. Your perspectives have made it the best version of itself. And family and friends have been indispensable throughout: the Nancollas and Burness families, Jenny and Harry Morris, Kate and Mark Blatchford, Bryan and Mary Jones, Mo O'Mahony, Yan Yates, Sam Brown, Tara Minton, Conor O'Brien, Seth Woodmansterne, the Parkinsons and the Moles. My thanks also to colleagues at the City of London Corporation, especially Jo Parker, Ben Eley, Kathryn Stubbs and Gwyn Richards, for their help and encouragement.

Lastly (but certainly not least), my unending love and gratitude to Josa and Ida, buoyancies during the more difficult passages encountered, and always the brightest stars by which to steer.

Index

Index

Index

Index

Index

Index

Index